Adventures of a Computational Explorer

Stephen Wolfram

Adventures of a *Computational* *Explorer*

Adventures of a Computational Explorer
Copyright © 2019 Stephen Wolfram, LLC

Wolfram Media, Inc. | wolfram-media.com

ISBN 978-1-57955-026-4 (hardback)
ISBN 978-1-57955-027-1 (ebook)

Biography / Science

Library of Congress Cataloging-in-Publication Data

Wolfram, Stephen, author.
Adventures of a computational explorer / Stephen Wolfram.
First edition. Champaign, Illinois : Stephen Wolfram, LLC, [2019] Collection of essays the author has written
over the past dozen years for various occasions. LCCN 2019012752 (print) LCCN 2019016518 (ebook) ISBN
9781579550271 (ebook) ISBN 9781579550264 (hardcover : acid-free paper) LCSH: Computer science. Wolfram,
Stephen. Computer scientists-United States-Biography. LCC QA76.24 (ebook) LCC QA76.24.W65 2019
(print) DDC 004—dc23 LC record available at https://lccn.loc.gov/2019012752

Sources for photos and archival materials that are not from the author's collection or in the public domain:
pp. 1, 4, 20, 26: Paramount Pictures; pp. 4: Amy Adams, Denis Villeneuve; pp. 39: Keith Schengili-Roberts;
pp. 42: Clemens Schmillen, Pablo Gimenez; pp. 42, 43, 60, 66, 69: Getty Images; pp. 41: Ames Construction;
pp. 43: Eric Coqueugniot, CNRS; pp. 44, 63–72: NASA; pp. 44, 45: PBS-WTVP, Big Pacific; pp. 63: Pauli Rautakorpi;
pp. 66: Cosmosphere, Kansas; pp. 74: The Planetary Society; pp. 119: Centre for Computer History; pp. 120:
Berkeley Physics, McGraw Hill, 1964; pp. 121: *Nuclear Physics B*, 1976; pp. 122: Anthony Hearn; pp. 123: M. J.
G. Veltman; pp. 123: Computer History Museum; pp. 183: Twitch.tv; pp. 218: ETH-Bibliothek Zürich; pp. 224:
J. Mater. Sci., A. R. Kortan, H. S. Chen, J. M. Parsey et al., 1989; B. Dubost, J-M. Lang et al. *Nature* 324, 48–50,
1986; pp. 225: P. Guyot, *Nature* 326, 640–641, 1987; pp. 230: Yolanda Cipriano; pp.230–1: Paula Guerra; pp. 235:
Sit Kong Sang, art by Flávio Império; pp. 337, 343: Alyssa Adams; pp. 349: Jared Tarbell (CA Chain); Kristoffer
Myskja (hole-punch); Troika (cubes); Fabienne Serriere (scarf); Cam Fox (tea cozy); www.oneandother.io,
@oneandother.io (shirt); Jeff Cook (block); art by Sultra & Barthélémy, automata by Nazim Fatès (rug); Gavin
Smith (worksheets)

Printed by Friesens, Manitoba, Canada. ∞ Acid-free paper. First edition. First printing.

Contents

Preface

"You work so hard... but what do you do for fun?" people will ask me. Well, the fact is that I've tried to set up my life so that the things I work on are things I find fun. Most of those things are aligned with big initiatives of mine, and with products and companies and scientific theories that I've built over decades. But sometimes I work on things that just come up, and that for one reason or another I find interesting and fun.

This book is a collection of pieces I've written over the past dozen years on some of these things, and the adventures I've had around them. Most of the pieces I wrote in response to some particular situation or event. Their topics are diverse. But it's remarkable how connected they end up being. And at some level all of them reflect the paradigm for thinking that has defined much of my life.

It all centers around the idea of computation, and the generality of abstraction to which it leads. Whether I'm thinking about science, or technology, or philosophy, or art, the computational paradigm provides both an overall framework and specific facts that inform my thinking. And in a sense this book reflects the breadth of applicability of this computational paradigm.

But I suppose it also reflects something else that I've long cultivated in myself: a willingness and an interest in applying my ways of thinking to pretty much any topic. I sometimes imagine that I will have nothing much to add to some particular topic. But it's remarkable how often the computational paradigm—and my way of thinking about it—ends up providing a new and different insight, or an unexpected way forward.

I often urge people to "keep their thinking apparatus engaged" even when they're faced with issues that don't specifically seem to be in their domains of expertise. And I make a point of doing this myself. It helps that the computational paradigm is so broad. But even at a much more specific level I'm continually amazed by how much the things I've learned from science or language design or technology development or business actually do end up connecting to the issues that come up.

If there's one thing that I hope comes through from the pieces in this book it's how much fun it can be to figure things out, and to dive deep into understanding particular topics and questions. Sometimes there's a simple, superficial answer. But for me what's really exciting is the much more serious intellectual exploration that's involved in giving a proper, foundational answer. I always find it particularly fun when there's a very practical problem to solve, but to get to a good solution requires an adventure that takes one through deep, and often philosophical, issues.

Inevitably, this book reflects some of my personal journey. When I was young I thought my life would be all about making discoveries in specific areas of science. But what I've come to realize—particularly having embraced the computational paradigm—is that the same intellectual thought processes can be applied not just to what one thinks of as science, but to pretty much anything. And for me there's tremendous satisfaction in seeing how this works out.

Stephen Wolfram

Quick, How Might the Alien Spacecraft Work?

November 10, 2016

Connecting with Hollywood

"It's an interesting script" said someone on our PR team. It's pretty common for us to get requests from movie-makers about showing our graphics or posters or books in movies. But the request this time was different: could we urgently help make realistic screen displays for a big Hollywood science fiction movie that was just about to start shooting?

Well, in our company unusual issues eventually land in my inbox, and so it was with this one. Now it so happens that through some combination of relaxation and professional interest I've probably seen basically every mainstream science fiction movie that's appeared over the past few decades. But just based on the working title ("Story of Your Life") I wasn't even clear that this movie was science fiction, or what it was at all.

But then I heard that it was about first contact with aliens, and so I said, "sure, I'll read the script". And, yes, it was an interesting script. Complicated, but interesting. I couldn't tell if the actual movie would be mostly science fiction or mostly a love story. But there were definitely interesting science-related themes in it—albeit mixed with things that didn't seem to make sense, and a liberal sprinkling of minor science gaffes.

When I watch science fiction movies I have to say I quite often cringe, thinking, "Someone's spent $100 million on this movie—and yet they've made some gratuitous science mistake that could have been fixed in an instant if they'd just asked the right person." So I decided that even though it was a very busy time for me, I should get involved in what's now called *Arrival* and personally try to give it the best science I could.

There are, I think, several reasons Hollywood movies often don't get as much science input as they should. The first is that movie-makers usually just aren't sensitive to the "science texture" of their movies. They can tell if things are out of whack at a human level, but they typically can't tell if something is scientifically off. Sometimes they'll get as far as calling a local university for help, but too often they're sent to a hyper-specialized academic who'll not-very-usefully tell them their whole story is wrong. Of course, to be fair, science content usually doesn't make or break movies. But I think having good science content—like, say, good set design—can help elevate a good movie to greatness.

As a company we've had a certain amount of experience working with Hollywood, for example writing all the math for six seasons of the television show *Numb3rs*. I hadn't personally been involved—though I have quite a few science friends who've helped with movies. There's Jack Horner, who worked on *Jurassic Park*, and ended up (as he tells it) pretty much having all his paleontology theories in the movie, including ones that turned out to be wrong. And then there's Kip Thorne (famous for the recent triumph of detecting gravitational waves), who as a second career in his 80s was the original driving force behind *Interstellar*—and who made the original black hole visual effects with Mathematica. From an earlier era there was Marvin Minsky who consulted on AI for *2001: A Space Odyssey*, and Ed Fredkin who ended up as the model for the rather eccentric Dr. Falken in *WarGames*. And recently there was Manjul Bhargava, who for a decade shepherded what became *The Man Who Knew Infinity*, eventually carefully "watching the math" in weeks of editing sessions.

All of these people had gotten involved with movies much earlier in their production. But I figured that getting involved when the movie was about to start shooting at least had the advantage that one knew the movie was actually going to get made (and yes, there's often a remarkably high noise-to-signal ratio about such things in Hollywood). It also meant that my role was clear: all I could do was try to uptick and smooth out the science; it wasn't even worth thinking about changing anything significant in the plot.

The inspiration for the movie had come from an interesting 1998 short story by Ted Chiang. But it was a conceptually complicated story, riffing off a fairly technical idea in mathematical physics—and I wasn't alone in wondering how anyone could possibly make a movie out of it. Still, there it was, a 120-page script that basically did it, with some science from the original story, and quite a lot added, mostly still in a rather "lorem ipsum" state. And so I went to work, making comments, suggesting fixes, and so on.

A Few Weeks Later...

Cut to a few weeks later. My son Christopher and I arrive on the set of *Arrival* in Montreal. The latest X-Men movie is filming at a huge facility next door. *Arrival* is at a more modest facility. We get there when they're in the middle of filming a scene inside a helicopter. We can't see the actors, but we're watching on the "video village" monitor, along with a couple of producers and other people.

The first line I hear is "I've prepared a list of questions [for the aliens], *starting with some binary sequences…* ". And I'm like, "Wow, I suggested saying that! This is great!" But then there's another take. And a word changes. And then there are more takes. And, yes, the dialogue sounds smoother. But the meaning isn't right. And I'm realizing: this is more difficult than I thought. Lots of tradeoffs. Lots of complexity. (Happily, in the final movie, it ends up being a blend, with the right meaning, and sounding good.)

After a while there's a break in filming. We talk to Amy Adams, who plays a linguist assigned to communicate with the aliens. She's spent some time shadowing a local linguistics professor, and is keen to talk about the question of how much the language one uses determines how one thinks—which is a topic that as a computer-language designer I've long been interested in. But what the producers really want is for me to talk to Jeremy Renner, who plays a physicist in the movie. He's feeling out of sorts right then—so off we go to look at the "science tent" set they've built and think about what visuals will work with it.

Writing Code

The script made it clear that there were going to be lots of opportunities for interesting visuals. But much as I might have found it fun, I just didn't personally have the time to work on creating them. Fortunately, though, my son Christopher—who is a very fast and creative programmer—was interested in doing it. We'd hoped to just be able to ship him off to the set for a week or two, but it was decided he was still too young, so he started off working remotely.

His basic strategy was simple, just ask, "if we were doing this for real, what analysis and computations would we be doing?" We've got a list of alien landing sites; what's the pattern? We've got geometric data on the shape of the spacecraft; what's its significance? We've got alien "handwriting"; what does it mean?

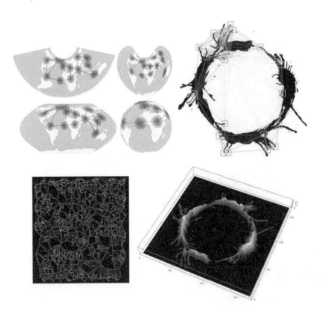

The movie-makers were giving Christopher raw data, just like in real life, and he was trying to analyze it. And he was turning each question that was asked into all sorts of Wolfram Language code and visualizations.

Christopher was well aware that code shown in movies often doesn't make sense (a favorite, regardless of context, seems to be the source code for nmap.c in Linux). But he wanted to create code that would make sense, and would actually do the analyses that would be going on in the movie.

```
In[ ]:= GeoGraphics[{Thickness[0.001],
        {Red, GeoPath /@ (List @@@ EdgeList[NearestNeighborGraph[landingSites, 3]])},
        Table[GeoDisk[#, n mi] & /@ landingSites, {n, 0, 1000, 250}], Red,
        GeoStyling[Opacity[1]], GeoDisk[#, 50 mi] & /@ landingSites},
       GeoRange → "World", GeoProjection → "WagnerII", GeoZoomLevel → 3]
```

Out[ ]=

```
In[ ]:= Module[{i = 🔵, corners}, corners = ImageCorners[i, 3, 0.1, 5];

  Show[{i, Graphics[{
    {Orange, Thickness[0.003], Outer[If[#1 === #2, {},
     {Opacity[3000 / EuclideanDistance[#1, #2] ^ 2],
     Line[{#1, #2}]}] &, corners, corners, 1]}, {EdgeForm[Green],
    FaceForm[], Rectangle[# - 10, # + 10] & /@ corners} }]}]]
```

In the final movie, the screen visuals are a mixture of ones Christopher created, ones derived from what he created, and ones that were put in separately. Occasionally one can see code. Like there's a nice shot of rearranging alien "handwriting," in which one sees a Wolfram Language notebook with rather elegant Wolfram Language code in it. And, yes, those lines of code actually do the transformation that's in the notebook. It's real stuff, with real computations being done.

A Theory of Interstellar Travel

When I first started looking at the script for the movie, I quickly realized that to make coherent suggestions I really needed to come up with a concrete theory for the science of what might be going on. Unfortunately there wasn't much time—and in the end I basically had just one evening to invent how interstellar space travel might work. Here's the beginning of what I wrote for the movie-makers about what I came up with that evening (to avoid spoilers I'm not showing more):

Science (Fiction) of Interstellar Spacecraft

Stephen Wolfram (May 23, 2015)

Introduction

I've never thought seriously about building an interstellar spacecraft before. But I've thought a lot about fundamental physics ... and, somewhat to my surprise, based on my current physics hypotheses, I think I just figured out a plausible scheme (though far from our current technology) for making an interstellar spacecraft...

Structure of Space

Nobody knows what the lowest level structure of space is. My current hypothesis is that it's ultimately a network of nodes, where all that's defined is connectivity. Space in the sense we know it emerges as a large-scale feature, just like the continuum properties of fluids (like water) emerge on a large scale, even though underneath they're made of lots of discrete molecules.

(With this setup, the network for example doesn't have a definite dimension (like 3); that emerges only as a large-scale average.)

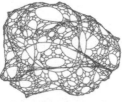

In this model, particles are just special regions in space (like vortices in water). (Einstein had a theory like this late in his life, though he couldn't make the details work.)

Motion happens a bit like with the structures here:

In a network, motion happens roughly by connections rearranging from being concentrated on one side of a particle (like an electron) to the other. (A bit like "swimming" through the network.)

When one lays the network out (e.g. in 3D space) most of the connections will be fairly local. But there will be a few long-range connections, roughly corresponding to quantum entanglement. These long-range connections in effect break out of the usual three dimensions of space.

If one could somehow get a large piece of network to be as disconnected as these long-range connections, then one would have a way to travel faster than light with respect to the large-scale average 3D space. (The effect is vaguely like a wormhole, or better a "warp bubble", but this gives an actual mechanism for how such things could be set up.)

Spaceship

The concept would be that somehow the outside of the ship defines a boundary layer that can be largely disconnected from the rest of the network. How would one achieve this?

Ordinary matter (with particles like electrons and protons) is composed of particular localized structures in the network. But one could imagine other structures in the network that would be less connected to the rest of the network.

This would be a form of matter that wouldn't necessarily visibly be made of standard elementary particles at all. It might be like a giant crystal formed directly from connections that make up space.

Most likely it would be dynamic. Also, unlike an ordinary crystal that has a regular periodic structure, it might have a much more complex (and continually changing) structure.

I would imagine that the skin of the ship could be a dynamic metamaterial whose detailed structure determines the interaction with the space outside.

Operation of the Spaceship

A possible scheme is to identify an entangled quantum particle that has come from a long way away (say a photon). Then somehow to take the lump of network that represents the spaceship, and find a way to attach all of it to the network connections that correspond to the entanglement.

It's like stuffing the whole spacecraft through the long-range network connection that's opened up by a single photon.

Obviously all these physics details weren't directly needed in the movie. But thinking them through was really useful in making consistent suggestions about the script. And they led to all sorts of science-fictiony ideas for dialogue. Here are a few of the ones that (probably for the better) didn't make it into the final script. "The whole ship goes through space like one giant quantum particle." "The aliens must directly manipulate the spacetime network at the Planck scale." "There's spacetime turbulence around the skin of the ship." "It's like the skin of the ship has an infinite number of types of atoms, not just the 115 elements we know" (that was going to be related to shining a monochromatic laser at the ship and seeing it come back looking like a rainbow). It's fun for an "actual scientist" like me to come up with stuff like this. It's kind of liberating. Especially since every one of these science-fictiony pieces of dialogue can lead one into a long, serious physics discussion.

For the movie, I wanted to have a particular theory for interstellar travel. And who knows, maybe one day in the distant future it'll turn out to be correct. But as of now, we certainly don't know. In fact, for all we know, there's just some simple "hack" in existing physics that'll immediately make interstellar travel possible. For example, there's even some work I did back in 1982 that implies that with standard quantum field theory one should, almost paradoxically, be able to continually extract "zero point energy" from the vacuum. And over the years, this basic mechanism has become what's probably the most quoted potential propulsion source for interstellar travel, even if I myself don't actually believe in it. (I think it takes idealizations of materials much too far.)

Maybe (as has been popular recently) there's a much more prosaic way to propel at least a tiny spacecraft, by pushing it to nearby stars with radiation pressure from a laser. Or maybe there's some way to do "black hole engineering" to set up appropriate distortions in spacetime, even in the standard Einsteinian theory of gravity. It's important to realize that even if (when?) we know the fundamental theory of physics, we still may not immediately be able to determine, for example, whether faster-than-light travel is possible in our universe.

Is there some way to set up some configuration of quantum fields and black holes and whatever so that things behave just so? Computational irreducibility (related to undecidability, Gödel's theorem, the Halting Problem, etc.) tells one that there's no upper bound on just how elaborate and difficult-to-set-up the configuration might need to be. And in the end one could use up all the computation that can be done in the history of the universe—and more—trying to invent the structure that's needed, and never know for sure if it's impossible.

What Are Physicists Like?

When we're visiting the set, we eventually meet up with Jeremy Renner. We find him sitting on the steps of his trailer smoking a cigarette, looking every bit the gritty action-adventurer that I realize I've seen him as in a bunch of movies. I wonder about the most efficient way to communicate what physicists are like. I figure I should just start talking about physics. So I start explaining the physics theories that are relevant to the movie. We're talking about space and time and quantum mechanics and faster-than-light travel and so on. I'm sprinkling in a few stories I heard from Richard Feynman about "doing physics in the field" on the Manhattan Project. It's an energetic discussion, and I'm wondering what mannerisms I'm displaying—that might or might not be typical of physicists. (I can't help remembering Oliver Sacks telling me how uncanny it was for him to see how many of his mannerisms Robin Williams had picked up for *Awakenings* after only a little exposure, so I'm wondering what Jeremy is going to pick up from me in these few hours.)

Jeremy is keen to understand how the science relates to the arc of the story for the movie, and what the aliens as well as humans must be feeling at different points. I try to talk about what it's like to figure stuff out in science. Then I realize the best thing is to actually show it a bit, by doing some livecoding. And it turns out that the way the script is written right then, Jeremy is actually supposed to be on camera using the Wolfram Language himself (just like—I'm happy to say—so many real-life physicists do).

Christopher shows some of the code he's written for the movie, and how the controls to make the dynamics work. Then we start talking about how one sets about figuring out the code. We do some preliminaries. Then we're off and running, doing livecoding. And here's the first example we make—based on the digits of pi that we'd been discussing in relation to SETI or *Contact* (the book version) or something:

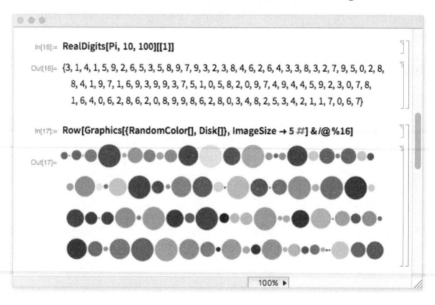

```
In[16]:= RealDigits[Pi, 10, 100][[1]]

Out[16]= {3, 1, 4, 1, 5, 9, 2, 6, 5, 3, 5, 8, 9, 7, 9, 3, 2, 3, 8, 4, 6, 2, 6, 4, 3, 3, 8, 3, 2, 7, 9, 5, 0, 2, 8,
        8, 4, 1, 9, 7, 1, 6, 9, 3, 9, 9, 3, 7, 5, 1, 0, 5, 8, 2, 0, 9, 7, 4, 9, 4, 4, 5, 9, 2, 3, 0, 7, 8,
        1, 6, 4, 0, 6, 2, 8, 6, 2, 0, 8, 9, 9, 8, 6, 2, 8, 0, 3, 4, 8, 2, 5, 3, 4, 2, 1, 1, 7, 0, 6, 7}

In[17]:= Row[Graphics[{RandomColor[], Disk[]}, ImageSize → 5 #] & /@ %16]
```

What to Say to the Aliens

Arrival is partly about interstellar travel. But it's much more about how we'd communicate with the aliens once they've showed up here. I've actually thought a lot about alien intelligence. But mostly I've thought about it in a more difficult case than in *Arrival*—where there are no aliens or spaceships in evidence, and where the only thing we have is some thin stream of data, say from a radio transmission, and where it's difficult even to know if what we've got should be considered evidence of "intelligence" at all (remember, for example, that it often seems that even the weather can be complex enough to seem like it "has a mind of its own").

But in *Arrival*, the aliens are right here. So then how should we start communicating with them? We need something universal that doesn't depend on the details of human language or human history. Well, OK, if you're right there with the aliens, there are physical objects to point to. (Yes, that assumes the aliens have some notion of discrete objects, rather than just a continuum, but by the time they've got spaceships and so on, that seems like a decently safe bet.) But what if you want to be more abstract?

Well, then there's always mathematics. But is mathematics actually universal? Does anyone who builds spaceships necessarily have to know about prime numbers, or integrals, or Fourier series? It's certainly true that in our human development of technology, those are things we've needed to understand. But are there other (and perhaps better) paths to technology? I think so.

For me, the most general form of abstraction that seems relevant to the actual operation of our universe is what we get by looking at the computational universe of possible programs. Mathematics as we've practiced it does show up there. But so do an infinite diversity of other abstract collections of rules. And what I realized a while back is that many of these are very relevant—and actually very good—for producing technology.

So, OK, if we look across the computational universe of possible programs, what might we pick out as reasonable universals to start an abstract discussion with aliens who've come to visit us?

Once one can point to discrete objects, one has the potential to start talking about numbers, first in unary, then perhaps in binary. Here's the beginning of a notebook I made about this for the movie. The words and code are for human consumption; for the aliens there'd just be "flash cards" of the main graphics:

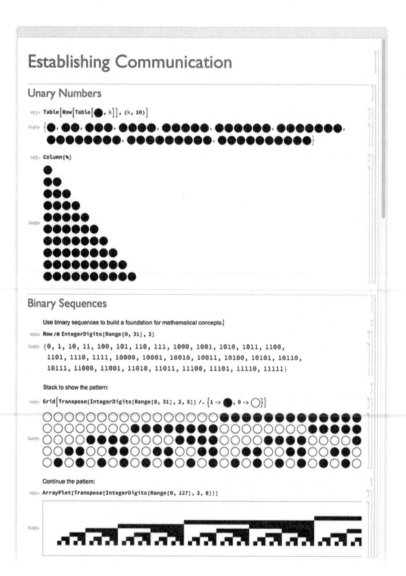

OK, so after basic numbers, and maybe some arithmetic, what's next? It's interesting to realize that even what we've discussed so far doesn't reflect the history of human mathematics: despite how fundamental they are (as well as their appearance in very old traditions like the *I Ching*) binary numbers only got popular quite recently—long after lots of much-harder-to-explain mathematical ideas.

We don't need to follow the history of human mathematics or science— or, for that matter, the order in which it's taught to humans, but we do need to find things that can be understood very directly—without outside knowledge or words. Things that for example we'd recognize if we just unearthed them without context in some archeological dig.

Well, it so happens that there's a class of computational systems that I've studied for decades that I think fit the bill remarkably well: cellular automata. They're based on simple rules that are easy to display visually. And they work by repeatedly applying these rules, and often generating complex patterns—that we now know can be used as the basis for all sorts of interesting technology.

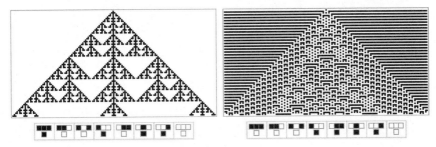

From looking at cellular automata one can actually start to build up a whole world view, or, as I called the book I wrote about such things, *A New Kind of Science*. But what if we want to communicate more traditional ideas in human science and mathematics? What should we do then?

 Maybe we could start by showing 2D geometrical figures. Gauss suggested back around 1820 that one could carve a picture of the standard visual for the Pythagorean theorem out of the Siberian forest, for aliens to see.

It's easy to get into trouble, though. We might think of showing Platonic solids. And, yes, 3D printouts should work. But 2D perspective renderings depend on a lot of detail on our particular visual systems. Networks are even worse: how are we to know that those lines joining nodes represent abstract connections?

One might think about logic: perhaps start showing the true theorems of logic. But how would one present them? Somehow one has to have a symbolic representation: textual, expression trees, or something. From what we know now about computational knowledge, logic isn't a particularly good global starting point for representing general concepts. But in the 1950s this wasn't clear, and there was a charming book (my copy of which wound up on the set of *Arrival*) that tried to build up a whole way to communicate with aliens using logic:

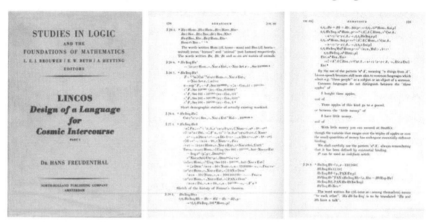

But what about things with numbers? In *Contact* (the movie), prime numbers are key. Well, despite their importance in the history of human mathematics, primes actually don't figure much in today's technology, and when they do (like in public-key cryptosystems) it usually seems somehow incidental that they're what's used.

In a radio signal, primes might at first seem like good "evidence for intelligence." But of course primes can be generated by programs—and actually by fairly simple ones, including for example cellular automata. And so if one sees a sequence of primes, it's not immediate evidence that there's a whole elaborate civilization behind it; it might just come from a simple program that somehow "arose naturally."

One can easily illustrate primes visually (not least as numbers of objects that can't be arranged in nontrivial rectangles). But going further with them seems to require concepts that can't be represented so directly.

It's awfully easy to fall into implicitly assuming a lot of human context. *Pioneer 10*—the human artifact that's gone further into interstellar space than any other (currently about 11 billion miles, which is about 0.05% of the distance to α Centauri)—provides one of my favorite examples. There's a plaque on that spacecraft that includes a representation of the wavelength of the 21-centimeter spectral line of hydrogen. Now the most obvious way to represent that would probably just be a line 21 cm long. But back in 1972 Carl Sagan and others decided to do something "more scientific," and instead made a schematic diagram of the quantum mechanical process leading to the spectral line. The problem is that this diagram relies on conventions from human textbooks—like using 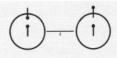 arrows to represent quantum spins—that really have nothing to do with the underlying concepts and are incredibly specific to the details of how science happened to develop for us humans.

But back to *Arrival*. To ask a question like "What is your purpose on Earth?" one has to go a lot further than just talking about things like binary sequences or cellular automata. It's a very interesting problem, and one that's strangely analogous to something that's becoming very important right now in the world: communicating with AIs, and defining what goals or purposes they should have (notably "be nice to the humans").

In a sense, AIs are a little like alien intelligences, right now, here on Earth. The only intelligence we really understand so far is human intelligence. But inevitably every example we see of it shares all the details of the human condition and of human history. So what is intelligence like when it doesn't share those details?

Well, one of the things that's emerged from basic science I've done is that there isn't really a bright line between the "intelligent" and the merely "computational." Things like cellular automata—or the weather—are doing things just as complex as our brains. But even if in some sense they're "thinking," they're not doing so in human-like ways. They don't share our context and our details.

But if we're going to "communicate" about things like purpose, we've got to find some way to align things. In the AI case, I've in fact been working on creating what I call a "symbolic discourse language" that's a way of expressing concepts that are important to us humans, and communicating them to AIs. There are short-term practical applications, like setting up smart contracts. And there are long-term goals, like defining some analog of a "constitution" for how AIs should generally behave.

Well, in communicating with aliens, we've got to build up a common "universal" language that allows us to express concepts that are important to us. That's not going to be easy. Human natural languages are based on the particulars of the human condition and the history of human civilization. And my symbolic discourse language is really just trying to capture things that are important to humans—not what might be important to aliens.

Of course, in *Arrival*, we already know that the aliens share some things with us. After all, like the monolith in *2001: A Space Odyssey*, even from their shape we recognize the aliens' spaceships as artifacts. They don't seem like weird meteorites or something; they seem like something that was made "on purpose."

But what purpose? Well, purpose is not really something that can be defined abstractly. It's really something that can be defined only relative to a whole historical and cultural framework. So to ask aliens what their purpose is, we first have to have them understand the historical and cultural framework in which we operate.

Somehow I wonder about the day when we'll have developed our AIs to the point where we can start asking them what their purpose is. At some level I think it's going to be disappointing. Because, as I've said, I don't think there's any meaningful abstract definition of purpose. So there's nothing "surprising" the AI will tell us. What it considers its purpose will just be a reflection of its detailed history and context. Which in the case of the AI—as its ultimate creators—we happen to have considerable control over.

For aliens, of course, it's a different story. But that's part of what *Arrival* is about.

The Movie Process

I've spent a lot of my life doing big projects—and I'm always curious how big projects of any kind are organized. When I see a movie I'm one of those people who sits through to the end of the credits. So it was pretty interesting for me to see the project of making a movie a little closer up in *Arrival*.

In terms of scale, making a movie like *Arrival* is a project of about the same size as releasing a major new version of the Wolfram Language. And it's clear there are some similarities—as well as lots of differences.

Both involve all sorts of ideas and creativity. Both involve pulling together lots of different kinds of skills. Both have to have everything fit together to make a coherent product in the end.

In some ways I think movie-makers have it easier than us software developers. After all, they just have to make one thing that people can watch. In software—and particularly in language design—we have to make something that different people can use in an infinite diversity of different ways, including ones we can't directly foresee. Of course, in software you always get to make new versions that incrementally improve things; in movies you just get one shot.

And in terms of human resources, there are definitely ways software has it easier than a movie like *Arrival*. Well-managed software development tends to have a somewhat steady rhythm, so one can have consistent work going on, with consistent teams, for years. In making a movie like *Arrival* one's usually bringing in a whole sequence of people—who might never even have met before—each for a very short time. To me, it's amazing this can work at all. But I guess over the years many of the tasks in the movie industry have become standardized enough that someone can be there for a week or two and do something, then successfully hand it off to another person.

I've led a few dozen major software releases in my life. And one might think that by now I'd have got to the point where doing a software release would just be a calm and straightforward process. But it never is. Perhaps it's because we're always trying to do majorly new and innovative things. Or perhaps it's just the nature of such projects. But I've found that to get the project done to the quality level I want always requires a remarkable degree of personal intensity. Yes, at least in the case of our company, there are always extremely talented people working on the project. But somehow there are always things to do that nobody expected, and it takes a lot of energy, focus, and pushing to get them all together.

At times, I've imagined that the process might be a little like making a movie. And in fact in the early years of Mathematica, for example, we even used to have "software credits" that looked very much like movie credits—except that the categories of contributors were things that often had to be made up by me ("lead package developers," "expression formatting," "lead font designer," ...). But after a decade or so, recognizing the patchwork of contributions to different versions just became too complex, and so we had to give up on software credits. Still, for a while I thought we'd try having "wrap parties," just like for movies. But somehow when the scheduled party came around, there was always some critical software issue that had come up, and the key contributors couldn't come to the party because they were off fixing it.

Software development—or at least language development—also has some structural similarities to movie-making. One starts from a script—an overall specification of what one wants the finished product to be like. Then one actually tries to build it. Then, inevitably, at the end when one looks at what one has, one realizes one has to change the specification. In movies like *Arrival*, that's post-production. In software, it's more an iteration of the development process.

It was interesting to me to see how the script and the suggestions I made for it propagated through the making of *Arrival*. It reminded me quite a lot of how I, at least, do software design: everything kept on getting simpler. I'd suggest some detailed way to fix a piece of dialogue. "You shouldn't say [the Amy Adams character] flunked calculus; she's way too analytical for that." "You shouldn't say the spacecraft came a million light years; that's outside the galaxy; say a trillion miles instead." The changes would get made. But then things would get simpler, and the core idea would get communicated in some more minimal way. I didn't see all the steps (though that would have been interesting). But the results reminded me quite a lot of the process of software design I've done so many times—cut out any complexity one can, and make everything as clear and minimal as possible.

Can You Write a Whiteboard?

My contributions to *Arrival* were mostly concentrated around the time the movie was shooting early in the summer of 2015. And for almost a year all I heard was that the movie was "in post-production." But then suddenly in May of this year I get an email: could I urgently write a bunch of relevant physics on a whiteboard for the movie?

There was a scene with Amy Adams in front of a whiteboard, and somehow what was written on the whiteboard when the scene was shot was basic high-school-level physics—not the kind of top-of-the-line physics one would expect from people like the Jeremy Renner character in the movie.

Somewhat amusingly, I don't think I've ever written much on a whiteboard before. I've used computers for essentially all my work and presentations for more than 30 years, and before that the prevailing technologies were blackboards and overhead projector transparencies. Still, I duly got a whiteboard set up in my office, and got to work writing (in my now-very-rarely-used handwriting) some things I imagined a good physicist might think of if they were trying to understand an interstellar spacecraft that had just showed up.

Here's what I came up with. The big spaces on the whiteboard were there to make it easier to composite in Amy Adams (and particularly her hair) moving around in front of the whiteboard. (In the end, the whiteboard got rewritten yet again for the final movie, so what's here isn't in detail what's in the movie.)

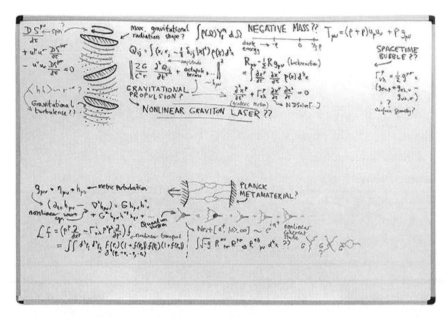

In writing the whiteboard, I imagined it as a place where the Jeremy Renner character or his colleagues would record notable ideas about the spacecraft, and formulas related to them. And after a little while, I ended up with quite a tale of physics fact and speculation.

Here's a key:

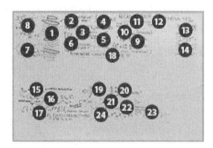

(1) Maybe the spacecraft has its strange (here, poorly drawn) rattleback-like shape because it spins as it travels, generating gravitational waves in spacetime in the process.

(2) Maybe the shape of the spacecraft is somehow optimized for producing a maximal intensity of some pattern of gravitational radiation.

(3) This is Einstein's original formula for the strength of gravitational radiation emitted by a changing mass distribution. Q_{ij} is the quadrupole moment of the distribution, computed from the integral shown.

(4) There are higher-order terms, that depend on higher-order multipole moments, computed by these integrals of the spacecraft mass density $\rho(\Omega)$ weighted by spherical harmonics.

(5) The gravitational waves would lead to a perturbation in the structure of spacetime, represented by the 4-dimensional tensor $h_{\mu\nu}$.

(6) Maybe the spacecraft somehow "swims" through spacetime, propelled by the effects of these gravitational waves.

(7) Maybe around the skin of the spacecraft, there's "gravitational turbulence" in the structure of spacetime, with power-law correlations like the turbulence one sees around objects moving in fluids. (Or maybe the spacecraft just "boils spacetime" around it...)

(8) This is the Papapetrou equation for how a spin tensor evolves in General Relativity, as a function of proper time τ.

(9) The equation of geodesic motion describing how things move in (potentially curved) spacetime. Γ is the Christoffel symbol determined by the structure of spacetime. And, yes, one can just go ahead and solve such equations using NDSolve in the Wolfram Language.

(10) Einstein's equation for the gravitational field produced by a moving mass (the field determines the motion of the mass, which in turn reacts back to change the field).

(11) A different idea is that the spacecraft might somehow have negative mass, or at least negative pressure. A photon gas has pressure $1/3\,\rho$; the most common version of dark energy would have pressure $-\rho$.

(12) The equation for the energy–momentum tensor, that specifies the combination of mass, pressure, and velocity that appears in relativistic computations for perfect fluids.

(13) Maybe the spacecraft represents a "bubble" in which the structure of spacetime is different. (The arrow pointed to a schematic spacecraft shape pre-drawn on the whiteboard.)

(14) Is there anything special about the Christoffel symbols ("coefficients of the connection on the tangent fiber bundle") for the shape of the spacecraft, as computed from its spatial metric tensor?

(15) A gravitational wave can be described as a perturbation in the metric of spacetime relative to flat background Minkowski space where Special Relativity operates.

(16) The equation for the propagation of a gravitational wave, taking into account the first few "nonlinear" effects of the wave on itself.

(17) The relativistic Boltzmann equation describing motion ("transport") and collision in a gas of Bose–Einstein particles like gravitons.

(18) A far-out idea: maybe there's a way of making a "laser" using gravitons rather than photons, and maybe that's how the spacecraft works.

(19) Lasers are a quantum phenomenon. This is a Feynman diagram of self-interaction of gravitons in a cavity. (Photons don't have these kinds of direct "nonlinear" self-interactions.)

(20) How might one make a mirror for gravitons? Maybe one can make a metamaterial with a carefully constructed microscopic structure all the way down to the Planck scale.

(21) Lasers involve coherent states made from superpositions of infinite numbers of photons, as formed by infinitely nested creation operators applied to the quantum field theoretic vacuum.

(22) There's a Feynman diagram for that: this is a Bethe–Salpeter-type self-consistent equation for a graviton bound state (which we don't know exists) that might be relevant to a graviton laser.

(23) Basic nonlinear interactions of gravitons in a perturbative approximation to quantum gravity.

(24) A possible correction term for the Einstein–Hilbert action of General Relativity from quantum effects.

Eek, I can see how these explanations might seem like they're in an alien language themselves! Still, they're actually fairly tame compared to "full physics-speak." But let me explain a bit of the "physics story" on the whiteboard.

It starts from an obvious feature of the spacecraft: its rather unusual, asymmetrical shape. It looks a bit like one of those rattleback tops that one can start spinning one way, but then it changes direction. So I thought: maybe the spacecraft spins around. Well, any massive (non-spherical) object spinning around will produce gravitational waves. Usually they're absurdly too weak to detect, but if the object is sufficiently massive or spins sufficiently rapidly, they can be substantial. And indeed, late last year, after a 30-year odyssey, gravitational waves from two black holes spinning around and merging were detected—and they were intense enough to detect from a third of the way across the universe. (Accelerating masses effectively generate gravitational waves like accelerating electric charges generate electromagnetic waves.)

OK, so let's imagine the spacecraft somehow spins rapidly enough to generate lots of gravitational waves. And what if we could somehow confine those gravitational waves in a small region, maybe even by using the motion of the spacecraft itself? Well, then the waves would interfere with themselves. But what if the waves got coherently amplified, like in a laser? Well, then the waves would get stronger, and they'd inevitably start having a big effect on the motion of the spacecraft—like perhaps pushing it through spacetime.

But why should the gravitational waves get amplified? In an ordinary laser that uses photons ("particles of light"), one basically needs to continually make new photons by pumping energy into a material. Photons are so-called Bose–Einstein particles ("bosons") which means that they tend to all "do the same thing"—which is why the light in a laser comes out as a coherent wave. (Electrons are fermions, which means that they try never to do the same thing, leading to the Exclusion Principle that's crucial in making matter stable, etc.)

Just as light waves can be thought of as made up of photons, so gravitational waves can most likely be thought of as made up of gravitons (though, to be fair, we don't yet have any fully consistent theory of gravitons). Photons don't interact directly with each other—basically because photons interact with things like electrons that have electric charge, but photons themselves don't have electric charge. Gravitons, on the other hand, do interact directly with each other—basically because they interact with things that have any kind of energy, and they themselves can have energy.

These kinds of nonlinear interactions can have wild effects. For example, gluons in QCD have nonlinear interactions that have the effect of keeping them permanently confined inside the particles like protons that they keep "glued" together. It's not at all clear what non-linear interactions of gravitons might do. The idea here is that perhaps they'd lead to some kind of self-sustaining "graviton laser."

The formulas at the top of the whiteboard are basically about the generation and effects of gravitational waves. The ones at the bottom are mostly about gravitons and their interactions. The formulas at the top are basically all associated with Einstein's General Theory of Relativity (which for 100 years has been the theory of gravity used in physics). The formulas at the bottom give a mixture of classical and quantum approaches to gravitons and their interactions. The diagrams are so-called Feynman diagrams, in which wavy lines schematically represent gravitons propagating through spacetime.

I have no real idea if a "graviton laser" is possible, or how it would work. But in an ordinary photon laser, the photons always effectively bounce around inside some kind of cavity whose walls act as mirrors. Unfortunately, however, we don't know how to make a graviton mirror—just like we don't know any way of making something that will shield a gravitational field (well, dark matter sort of would, if it actually exists). For the whiteboard, I made the speculation that perhaps there's some weird way of making a "metamaterial" down at the Planck scale of 10^{-34} meters (where quantum effects in gravity basically have to become important) that could act as a graviton mirror. (Another possibility is that a graviton laser could work more like a free-electron laser without a cavity as such.)

Now, remember, my idea with the whiteboard was to write what I thought a typical good physicist, say plucked from a government lab, might think about if confronted with the situation in the movie. It's more "conventional" than the theory I personally came up with for how to make an interstellar spacecraft. But that's because my theory depends on a bunch of my own ideas about how fundamental physics works, that aren't yet mainstream in the physics community.

What's the correct theory of interstellar travel? Needless to say, I don't know. I'd be amazed if either the main theory I invented for the movie or the theory on the whiteboard were correct as they stand. But who knows? And of course it'd be extremely helpful if some aliens showed up in interstellar spaceships to even show us that interstellar travel is possible...

What Is Your Purpose on Earth?

If aliens show up on Earth, one of the obvious big questions is: why are you here? What is your purpose? It's something the characters in *Arrival* talk about a lot. And when Christopher and I were visiting the set we were asked to make a list of possible answers, that could be put on a whiteboard or a clipboard. Here's what we came up with:

As I mentioned before, the whole notion of purpose is something that's very tied into cultural and other contexts. And it's interesting to think about what purposes one would have put on this list at different times in human history. It's also interesting to imagine what purposes humans—or AIs—might give for doing things in the future. Perhaps I'm too pessimistic but I rather expect that for future humans, AIs, and aliens, the answer will very often be something out there in the computational universe of possibilities—that we today aren't even close to having words or concepts for.

And Now It's a Movie...

The movie came together really well, the early responses look great... and it's fun to see things like this (yes, that's Christopher's code):

It's been interesting and stimulating to be involved with *Arrival*. It's let me understand a little more about just what's involved in creating all those movies I see—and what it takes to merge science with compelling fiction. It's also led me to ask some science questions beyond any I've asked before—but that relate to all sorts of things I'm interested in.

But through all of this, I can't help wondering: "What if it was real, and aliens did arrive on Earth?" I'd like to think that being involved with *Arrival* has made me a little more prepared for that. And certainly if their spaceships do happen to look like giant black rattlebacks, we'll even already have some nice Wolfram Language code for that...

My Hobby: Hunting for Our Universe

September 11, 2007

I don't have much time for hobbies these days, but occasionally I get to indulge a bit. A few days ago I did a videoconference talking about one of my favorite hobbies: hunting for the fundamental laws of physics.

Physics was my first field (in fact, I became a card-carrying physicist when I was a teenager). And as it happens, the talk I just gave (for the European Network on Random Geometry) was organized by one of my old physics collaborators.

Physicists often like to think that they're dealing with the most fundamental kinds of questions in science. But actually, what I realized back in 1981 or so is that there's a whole layer underneath. There's not just our own physical universe to think about, but the whole universe of possible universes. If one's going to do theoretical science, one had better be dealing with some kind of definite rules. But the question is: what rules?

Nowadays we have a great way to parametrize possible rules: as possible computer programs. And I've built a whole science out of studying the universe of possible programs—and have discovered that even very simple ones can generate all sorts of rich and complex behavior. That's turned out to be relevant in modeling all sorts of systems in the physical and biological and social sciences, and in discovering interesting technology, and so on. But here's my big hobby question: what about our physical universe? Could it be operating according to one of these simple rules?

If the rules are simple enough, one might be able to do something that seems quite outrageous: just search the universe of all possible rules, and find our own physical universe.

It's certainly not obvious that our universe has simple rules at all. In fact, looking at all the complex stuff that goes on in the universe, one might think that the rules couldn't be terribly simple. Of course, as early theologians pointed out, the universe clearly has some order, some "design." It could be that every particle in the universe has its own separate rule, but in reality things are much simpler than that.

But just how simple? A thousand lines of Mathematica code? A million lines? Or, say, three lines? If it's small enough, we really should be able to find it just by searching. And I think it'd be embarrassing if our universe is out there, findable by today's technology, and we didn't even try.

Of course, that's not at all how most of today's physicists like to think. They like to imagine that by pure thought they can somehow construct the laws for the universe—like universe engineers. The physicists at my recent videoconference are a little closer to my point of view, though the methodology and technicalities of what I'm doing are still pretty alien to them.

But OK, so if there's a simple rule for the universe, what might it actually be like? I've done a lot of work on this, and written quite a lot about it. One important thing to realize is that if the rule is simple, it almost inevitably won't explicitly show anything familiar from ordinary everyday physics. Because in a really small rule, there just isn't room to fit an explicit "3D" for the effective dimension of space, or the explicit masses of one's favorite particles. In fact, there almost certainly isn't even room to fit an explicit notion of space, or of time.

So in a sense we have to go below space and time—to more fundamental primitives. So what might these be? There are undoubtedly many ways to formulate them. But I think most of the promising possibilities are ultimately equivalent to networks like this:

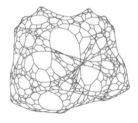

There's no "space" here—just a bunch of points, connected in a certain way. But I think it's a little like, say, a liquid: even though at the lowest level there are just a bunch of molecules bouncing around, on a large enough scale a continuum structure emerges.

Normally in physics one thinks of space as some kind of background, in which matter and particles and so on separately exist. But I suspect it's really more integrated: that everything is "just space", with the particles being something like special little lumps of connectivity in the network corresponding to space.

In his later years, Albert Einstein actually tried hard to construct models for physics a bit like this, in which everything emerged from space. But he had to use continuum equations as his "primitives", and he could never make it work.

Many years later, there are a certain number of physicists (many of whom were at my videoconference) who think about networks that might represent space. They haven't quite reached the level of abstractness that I'm at. They still tend to imagine that the points in the network have actual defined positions in some background space—or at least that there's some topology of faces defined. I'm operating at a more abstract level: all that's defined is the combinatorics of connections. Of course, one can always make a picture using GraphPlot or GraphPlot3D. But the details of that picture are quite arbitrary.

What's interesting, though, is that when a network gets big enough, its combinatorics alone can in effect define a correspondence with ordinary space. It doesn't always work. In fact, most networks (like the last two below) don't correspond to manifolds like 3D space. But some do. And I suspect our universe is one of them.

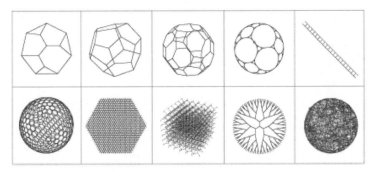

But having space isn't really enough. There's also time. Current physics tends to say that time is just like space—just another dimension. That's of course very different from the way it works in programs. In programs, moving in space might correspond to looking at another part of the data, but moving in time requires executing the program.

For networks, pretty much the most general kind of program is one that takes a piece of network with one structure, and replaces it with another.

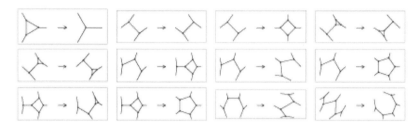

Often there'll be many different ways to apply rules like that to a particular network. And in general each possible sequence of rule applications might correspond to a "different branch of time." But it turns out that if one thinks about an entity inside the network (like us in the universe), then the only aspect of applying the rules that we can ever perceive is their "causal network": the network that says what "updating event" influences what other one.

Well, here's an important thing: there exist rules which have the property that whatever order they're applied in, they always give the same causal network.

And now there's a big fact: these causal invariant rules not only imply that there's just a single perceived thread of time in the universe; they also imply the particular relation of space and time that is Special Relativity.

Actually, there's even more than that. If the microscopic updatings of the underlying network end up being random enough, then it turns out that if the network succeeds in corresponding in the limit to a finite dimensional space, then this space must satisfy Einstein's Equations of General Relativity. It's again a little like what happens with fluids. If the microscopic interactions between molecules are random enough, but satisfy number and momentum conservation, then it follows that the overall continuum fluid must satisfy the standard Navier–Stokes equations.

But now we're deriving something like that for the universe: we're saying that these networks with almost nothing "built in" somehow generate behavior that corresponds to gravitation in physics.

This is all spelled out in the *NKS* book. And many physicists have certainly read that part of the book. But somehow every time I actually describe this (as I did a few days ago), there's a certain amazement. Special and General Relativity are things that physicists normally assume are built into theories right from the beginning, almost as axioms (or at least, in the case of string theory, as consistency conditions). The idea that they could emerge from something more fundamental is pretty alien.

The alien feeling doesn't stop there. Another thing that seems alien is the idea that our whole universe and its complete history could be generated just by starting with some particular small network, then applying definite rules.

For the past 75+ years, quantum mechanics has been the pride of physics, and it seems to suggest that this kind of deterministic thinking just can't be correct. It's a slightly long story (often still misunderstood by physicists), but between the arbitrariness of updating orders that produce a given causal network, and the fact that in a network one doesn't just have something like local 3D space, it looks as if one automatically starts to get a lot of the core phenomena of quantum mechanics—even from what's in effect a deterministic underlying model.

OK, but what is *the* rule for our universe? I don't know yet. Searching for it isn't easy. One tries a sequence of different possibilities. Then one runs each one. Then the question is: has one found our universe?

Well, sometimes it's easy to tell. Sometimes one's candidate universe disappears after a tiny amount of time. Or has some bizarre exponential version of space in which nothing can ever interact with anything else. Or some other pathology.

But the difficult cases are when what happens is more complicated. One starts one's candidate universe off. And it grows to millions or billions of nodes. And one can't see what it's doing. One uses GraphPlot. And lots of fancy analysis techniques. But all one can tell is that it's bubbling around, doing something complicated. Has one caught our universe, or not?

Well, here's the problem: one of the discoveries of *NKS* is a phenomenon I call computational irreducibility—which says that many systems that appear complex will have behavior that can never be "reduced" in general to a simpler computation.

It's inevitable that at some level our universe will have this property. But what we have to hope is that a candidate universe that we "catch in our net" will have enough reducibility that we can tell that it really is our universe.

What we've been doing for the past few years is to try to build technology for "universe identification." It's not at all trivial. In effect what we're trying to do is to build a system that can automatically recapitulate the whole history of physics—in a millisecond or something. We need to be able to take what we observe in our candidate universe, and somehow establish what its effective physical laws are, and see whether they correspond to our universe.

Of course, it's somehow more like mathematics than traditional physics. Because in a sense we have the underlying "axioms," and we're trying to see what laws they imply, rather than having to base everything on pure experiment.

There's an analogy that I find useful. When I was working on the *NKS* book, I wanted to understand some things about the foundations of mathematics. In particular, I wanted to know just where the mathematics that we do lies within the universe of all possible mathematics. So I started enumerating axiom systems, and trying to discover where in the space of possible axiom systems our familiar areas of mathematics show up.

One might think this was crazy—like searching for our universe in the space of possible universes. But *NKS* suggests it's not. Because it suggests that systems with simple rules can have the richness of anything.

And indeed, when I searched, for example, for Boolean algebra (logic), I did indeed find a tiny axiom system for it: it turned out to be about the 50,000th axiom system in the enumeration I used. Proving that it was correct took all sorts of fancy automated-theorem-proving technology—though I'm happy to say that in Mathematica, FullSimplify can just do it!

I think it's going to work a bit like this for the universe. It's going to take a lot of effort—and a little luck—to avoid the long arm of computational irreducibility. But the hope is that we'll be able to do it.

Physicists at the videoconference were very curious about whether I had candidate universes yet. The answer is yes. But I have no idea yet just how difficult they'll be to analyze.

A good friend of mine has kept on encouraging me not to throw away any even vaguely plausible universes—even if we can show that they're not our universe. He thinks that alternate universes have to be good for something.

I certainly think it'll be an interesting—almost metaphysical—moment if we finally have a simple rule which we can tell is our universe. And we'll be able to know that our particular universe is number such-and-such in the enumeration of all possible universes. It's a sort of Copernican moment: we'll get to know just how special or not our universe is.

Something I wonder is just how to think about whatever the answer turns out to be. It somehow reminds me of situations from earlier in the history of science. Newton figured out about motion of the planets, but couldn't imagine anything but a supernatural being first setting them in motion. Darwin figured out about biological evolution, but couldn't imagine how the first living cell came to be. We may have the rule for the universe, but it's something quite different to understand why it's that rule and not another.

Universe hunting is a very technology-intensive business. Over the years, I've gradually been building up the technology I think is needed—and quite a bit of it is showing up in strange corners of Mathematica. But I think it's going to be a while longer before there are more results. And before we can put "Our Universe" as a Demonstration in the Wolfram Demonstrations Project. And before we can take our new ParticleData computable data collection and derive every number in it.

But universe hunting is a good hobby.

Showing Off to the Universe: Beacons for the Afterlife of Our Civilization

January 25, 2018

The Nature of the Problem

Let's say we had a way to distribute beacons around our solar system (or beyond) that could survive for billions of years, recording what our civilization has achieved. What should they be like?

It's easy to come up with what I consider to be sophomoric answers. But in reality I think this is a deep—and in some ways unsolvable—philosophical problem, that's connected to fundamental issues about knowledge, communication, and meaning.

Still, a friend of mine recently started a serious effort to build little quartz disks, etc., and have them hitch rides on spacecraft, to be deposited around the solar system. At first I argued that it was all a bit futile, but eventually I agreed to be an advisor to the project, and at least try to figure out what to do to the extent we can.

But, OK, so what's the problem? Basically it's about communicating meaning or knowledge outside of our current cultural and intellectual context. We just have to think about archaeology to know this is hard. What exactly was some arrangement of stones from a few thousand years ago for? Sometimes we can pretty much tell, because it's close to something in our current culture. But a lot of the time it's really hard to tell.

OK, but what are the potential use cases for our beacons? One might be to back up human knowledge so things could be restarted even if something goes awfully wrong with our current terrestrial civilization. And of course historically it was very fortunate that we had all those texts from antiquity when things in Europe restarted during the Renaissance. But part of what made this possible was that there had been a continuous tradition of languages like Latin and Greek—not to mention that it was humans that were both the creators and consumers of the material.

But what if the consumers of the beacons we plan to spread around the solar system are aliens, with no historical connection to us? Well, then it's a much harder problem.

In the past, when people have thought about this, there's been a tendency to say, "Just show them math: it's universal, and it'll impress them!" But actually, I think neither claim about math is really true.

To understand this, we have to dive a little into some basic science that I happen to have spent many years working on. The reason people think math is a candidate for universal communication is that its constructs seem precise, and that at least here on Earth there's only one (extant) version of it, so it seems definable without cultural references. But if one actually starts trying to work out how to communicate about current math without any assumptions (as, for example, I did as part of consulting on the *Arrival* movie), one quickly discovers that one really has to go "below math" to get to computational processes with simpler rules.

And (as seems to happen with great regularity, at least to me) one obvious place one lands is with cellular automata. It's easy to show an elaborate pattern that's created according to simple, well-defined rules:

But here's the problem: there are plenty of physical systems that basically operate according to rules like these, and produce similarly elaborate patterns. So if this is supposed to show the impressive achievement of our civilization, it fails.

OK, but surely there must be something we can show that makes it clear that we've got some special spark of intelligence. I certainly always assumed there was. But one of the things that's come out of the basic science I've done is what I called the Principle of Computational Equivalence, that basically says that once one's gotten beyond a very basic level, every system will show behavior that's equivalent in the sophistication of the computation it exhibits.

So although we're very proud of our brains, and our computers, and our mathematics, they're ultimately not going to be able to produce anything that's beyond what simple programs like cellular automata—or, for that matter, "naturally occurring" physical systems—can produce. So when we make an offhand comment like "the weather has a mind of its own", it's not so silly: the fluid dynamic processes that lead to the weather are computationally equivalent to the processes that, for example, go on in our brains.

It's a natural human tendency at this point to protest that surely there must be something special about us, and everything we've achieved with our civilization. People may say, for example, that there's no meaning and no purpose to what the weather does. Of course, we can certainly attribute such things to it ("it's trying to equalize temperatures between here and there", etc.), and without some larger cultural story there's no meaningful way to say if they're "really there" or not.

OK, so if showing a sophisticated computation isn't going to communicate what's special about us and our civilization, what is? The answer is in the end details. Sophisticated computation is ubiquitous in our universe. But what's inevitably special about us are the details of our history and what we care about.

We're learning the same thing as we watch the progress of artificial intelligence. Increasingly, we can automate the things we humans can do—even ones that involve reasoning, or judgment, or creativity. But what we (essentially by definition) can't automate is defining what we want to do, and what our goals are. For these are intimately connected to the details of our biological existence, and the history of our civilization—which are exactly what's special about us.

But, OK, how can we communicate these things? Well, it's hard. Because—needless to say—they're tied into aspects of us that are special, and that won't necessarily be shared with whatever we're trying to communicate with.

At the end of the day, though, we've got a project that's going to launch beacons on spacecraft. So what's the best thing to put on them? I've spent a significant part of my life building what's now the Wolfram Language, whose core purpose is to provide a precise language for communicating knowledge that our civilization has accumulated in a way that both us humans and computers can understand. So perhaps this—and my experience with it—can help. But first, we should talk about history to get an idea of what has and hasn't worked in the past.

Lessons from the Past

A few years ago I was visiting a museum and looking at little wooden models of life in ancient Egypt that had been buried with some king several millennia ago. "How sad," I thought. "They imagined this would help them in the afterlife. But it didn't work; instead it just ended up in a museum." But then it struck me: "No, it did work! This is their 'afterlife'!" And they successfully transmitted some essence of their life to a world far beyond their own.

Of course, when we look at these models, it helps that a lot of what's in them is familiar from modern times. Cows. A boat with oars. Scrolls. But some isn't that familiar. What are those weird things at the ends of the boat, for example? What's the purpose of those? What are they for? And here begins the challenge—of trying to understand without shared context.

I happened last summer to visit an archaeological site in Peru named Caral, that has all sorts of stone structures built more than 4000 years ago. It was pretty obvious what some of the structures were for. But others I couldn't figure out. So I kept on asking our guide. And almost always the answer was the same: "it was for ceremonial purposes."

Immediately I started thinking about modern structures. Yes, there are monuments and public artworks. But there are also skyscrapers, stadiums, cathedrals, canals, freeway interchanges, and much more. And people have certain almost-ritual practices in interacting with these structures. But in the context of modern society, we would hardly call them "ceremonial": we think of each type of structure as having a definite purpose which we can describe. But that description inevitably involves a considerable depth of cultural context.

When I was growing up in England, I went wandering around in woods near where I lived—and came across all sorts of pits and berms and other earthworks. I asked people what they were. Some said they were ancient fortifications; some said at least the pits were from bombs dropped in World War II. And who knows: maybe instead they were created by some process of erosion having nothing to do with people.

Almost exactly 50 years ago, as a young child vacationing in Sicily, I picked up this object on a beach:

Being very curious what it was, I took it to my local archaeology museum. "You've come to the wrong place, young man," they said, "it's obviously a natural object." So off I went to a natural history museum, only to be greeted with "Sorry, it's not for us; it's an artifact." And from then until now the mystery has remained (though with modern materials analysis techniques it could perhaps be resolved—and I obviously should do it!)

There are so many cases where it's hard to tell if something is an artifact or not. Consider all the structures we've built on Earth. Back when I was writing *A New Kind of Science*, I asked some astronauts what the most obvious manmade structure they noticed from space was. It wasn't anything like the Great Wall of China (which is actually hard to see). Instead, they said it was a line across the Great Salt Lake in Utah (actually a 30-mile-long railroad causeway built in 1959, with algae that happen to have varying colors on the two sides):

Then there was the 12-mile-diameter circle in New Zealand, the 30-mile one in Mauritania, and the 40-mile one in Quebec (with a certain *Arrival* heptapod calligraphy look):

Which were artifacts? This was before the web, so we had to contact people to find out. A New Zealand government researcher told us not to make the mistake of thinking their circle followed the shape of the cone volcano at its center. "The truth is, alas, much more prosaic," he said: it's the border of a national park, with trees cut outside only, i.e. an artifact. The other circles, however, had nothing to do with humans.

(It's fun to look for evidence of humans visible from space. Like the grids of lights at night in Kansas, or lines of lights across Kazakhstan. And in recent years, there's the seven-mile-long palm tree rendering in Dubai. And, on the flip side, people have also tried to look for what might be "archaeological structures" in high-resolution satellite images of the Moon.)

But, OK, let's come back to the question of what things mean. In a cave painting from 7000 years ago, we can recognize shapes of animals, and hand stencils that we can see were made with hands. But what do the configurations of these things mean? Realistically at this point we have no serious idea.

Maybe it's easier if we look at things that are more "mathematical"-like. In the 1990s I did a worldwide hunt for early examples of complex but structured patterns. I found all sorts of interesting things (such as mosaics supposedly made by Gilgamesh, from 3000 BC—and the earliest fractals, from 1210 AD). Most of the time I could tell what rules were used to make the patterns—though I could not tell what "meaning" the patterns were supposed to convey, or whether, instead, they were "merely ornamental."

The last pattern above, though, had me very puzzled for a while. Is it a cellular automaton being constructed back in the 1300s? Or something from number theory? Well, no, in the end it turns out it's a rendering of a list of 62 attributes of Allah from the Koran, in a special square form of Arabic calligraphy constructed like this:

About a decade ago, I learned about a pattern from 11,000 years ago, on a wall in Aleppo, Syria (one hopes it's still intact there). What is this? Math? Music? Map? Decoration? Digitally encoded data? We pretty much have no idea.

I could go on giving examples. Lots of times people have said, "If one sees such-and-such, then it must have been made for a purpose." The philosopher Immanuel Kant offered the opinion that if one saw a regular hexagon drawn in the sand, one could only imagine a "rational cause" for it. I used to think of this whenever I saw hexagonal patterns formed in rocks. And a few years ago I heard about hexagons in sand, produced purely by the action of wind. But the biggest hexagon I know is the storm pattern around the north pole of Saturn—that presumably wasn't in any usual sense "put there for a purpose":

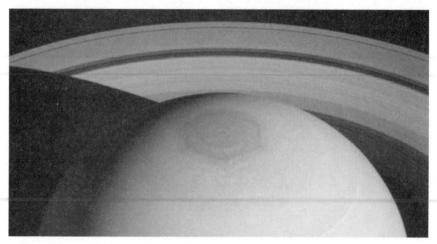

In 1899 Nikola Tesla picked up all sorts of elaborate and strange-sounding radio emissions, often a little reminiscent of Morse code. He knew they weren't of human origin, so his immediate conclusion was that they must be radio messages from the inhabitants of Mars. Needless to say, they're not. And instead, they're just the result of physical processes in the Earth's ionosphere and magnetosphere.

But here's the ironic thing: they often sound bizarrely similar to whale songs! And, yes, whale songs have all sorts of elaborate rhyme-like and other features that remind us of languages. But we still don't really know if they're actually for "communication," or just for "decoration" or "play."

One might imagine that with modern machine learning and with enough data one should be able to train a translator for "talking to animals." And no doubt that'd be easy enough for "are you happy?" or "are you hungry?" But what about more sophisticated things? Say the kind of things we want to communicate to aliens?

I think it'd be very challenging. Because even if animals live in the same environment as us, it's very unclear how they think about things. And it doesn't help that even their experience of the world may be quite different—emphasizing for example smell rather than sight, and so on.

Animals can of course make "artifacts" too. Like this arrangement of sand produced over the course of a week or so by a little puffer fish:

But what is this? What does it mean? Should we think of this "pisci-fact" as some great achievement of puffer fish civilization, that should be celebrated throughout the solar system?

Surely not, one might say. Because even though it looks complex—and even "artistic" (a bit like bird songs have features of music)—we can imagine that one day we'd be able to decode the neural pathways in the brain of the puffer fish that lead it to make this. But so what? We'll also one day be able to know the neural pathways in humans that lead them to build cathedrals—or try to plant beacons around the solar system.

Aliens and the Philosophy of Purpose

There's a thought experiment I've long found useful. Imagine a very advanced civilization, that's able to move things like stars and planets around at will. What arrangement would they put them in?

Maybe they'd want to make a "beacon of purpose." And maybe—like Kant—one could think that would be achievable by setting up some "recognizable" geometric pattern. Like how about an equilateral triangle? But no, that won't do. Because for example the Trojan asteroids actually form an equilateral triangle with Jupiter and the Sun already, just as a result of physics.

And pretty soon one realizes that there's actually nothing the aliens could do to "prove their purpose." The configuration of stars in the sky may look kind of random to us (except, of course, that we still see constellations in it). But there's nothing to say that looked at in the right way it doesn't actually represent some grand purpose.

And here's the confusing part: there's a sense in which it does! Because, after all, just as a matter of physics, the configuration that occurs can be characterized as achieving the purpose of extremizing some quantity defined by the equations for matter and gravity and so on. Of course, one might say "that doesn't count; it's just physics." But our whole universe (including ourselves) operates according to physics. And so now we're back to discussing whether the extremization is "meaningful" or not.

We humans have definite ways to judge what's meaningful or not to us. And what it comes down to is whether we can "tell a story" that explains, in culturally meaningful terms, why we're doing something. Of course, the notion of purpose has evolved over the course of human history. Imagine trying to explain walking on a treadmill, or buying goods in a virtual world, or, for that matter, sending beacons out into the solar system—to the people thousands of years ago who created the structures from Peru that I showed earlier.

We're not familiar (except in mythology) with telling "culturally meaningful stories" about the world of stars and planets. And in the past we might have imagined that somehow whatever stories we could tell would inevitably be far less rich than the ones we can tell about our civilization. But this is where basic science I've done comes in. The Principle of Computational Equivalence says that this isn't true—and that in the end what goes on with stars and planets is just as rich as what goes on in our brains or our civilization.

In an effort to "show something interesting" to the universe, we might have thought that the best thing to do would be to present sophisticated abstract computational things. But that won't be useful. Because those abstract computational things are ubiquitous throughout the universe.

And instead, the "most interesting" thing we have is actually the specific and arbitrary details of our particular history. Of course, one might imagine that there could be some sophisticated thing out there in the universe that could look at how our history starts, and immediately be able to deduce everything about how it will play out. But a consequence of the Principle of Computational Equivalence is what I call computational irreducibility, which implies that there can be no general shortcut to history; to find how it plays out, one effectively just has to live through it—which certainly helps one feel better about the meaningfulness of life.

The Role of Language

OK, so let's say we want to explain our history. How can we do it? We can't show every detail of everything that's happened. Instead, we need to give a higher-level symbolic description, where we capture what's important while idealizing everything else away. Of course, "what's important" depends on who's looking at it.

We might say, "Let's show a picture." But then we have to start talking about how to make the picture out of pixels at a certain resolution, how to represent colors, say with RGB—not to mention discussing how things might be imaged in 2D, compressed, etc. Across human history, we've had a decent record in having pictures remain at least somewhat comprehensible. But that's probably in no small part because our biologically determined visual systems have stayed the same.

(It's worth mentioning, though, that pictures can have features that are noticed only when they become "culturally absorbed." For example, the nested patterns from the 1200s that I showed earlier were reproduced but ignored in art history books for hundreds of years—until fractals became "a thing," and people had a way to talk about them.)

When it comes to communicating knowledge on a large scale, the only scheme we know (and maybe the only one that's possible) is to use language—in which essentially there's a set of symbolic constructs that can be arranged in an almost infinite number of ways to communicate different meanings.

It was presumably the introduction of language that allowed our species to begin accumulating knowledge from one generation to the next, and eventually to develop civilization as we know it. So it makes sense that language should be at the center of how we might communicate the story of what we've achieved.

And indeed if we look at human history, the cultures we know the most about are precisely those with records in written language that we've been able to read. If the structures in Caral had inscriptions, then (assuming we could read them) we'd have a much better chance of knowing what the structures were for.

There've been languages like Latin, Greek, Hebrew, Sanskrit, and Chinese that have been continuously used (or at least known) for thousands of years—and that we're readily able to translate. But in cases like Egyptian hieroglyphs, Babylonian cuneiform, Linear B, or Mayan, the thread of usage was broken, and it took heroic efforts to decipher them (and often the luck of finding something like the Rosetta Stone). And in fact today there are still plenty of languages—like Linear A, Etruscan, Rongorongo, Zapotec, and the Indus script—that have simply never been deciphered.

Then there are cases where it's not even clear whether something represents a language. An example is the quipus of Peru—that presumably recorded "data" of some kind, but that might or might not have recorded something we'd usually call a language:

Math to the Rescue?

OK, but with all our abstract knowledge about mathematics, and computation, and so on, surely we can now invent a "universal language" that can be universally understood. Well, we can certainly create a formal system—like a cellular automaton—that just consistently operates according to its own formal rules. But does this communicate anything?

In its actual operation, the system just does what it does. But where there's a choice is in what the actual system is, what rules it uses, and what its initial conditions were. So if we were using cellular automata, we could for example decide that these particular ones are the ones we want to show:

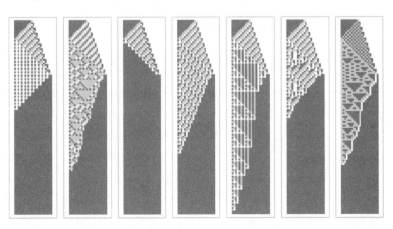

What are we communicating here? Each rule has all sorts of detailed properties and behavior. But as a human you might say: "Aha, I see that all these rules double the length of their input; that's the point." But to be able to make that summary again requires a certain cultural context. Yes, with our human intellectual history, we have an easy way to talk about "doubling the length of their input." But with a different intellectual history, that might not be a feature we have a way to talk about, just as human art historians for centuries didn't have a way to talk about nested patterns.

Let's say we choose to concentrate on traditional math. We have the same situation there. Maybe we could present theorems in some abstract system. But for each theorem it's just, "OK, fine, with those rules, that follows—much like with those shapes of molecules, this is a way they can arrange in a crystal." And the only way one's really "communicating something" is in the decision of which theorems to show, or which axiom systems to use. But again, to interpret those choices inevitably requires cultural context.

One place where the formal meets the actual is in the construction of theoretical models for things. We've got some actual physical process, and then we've got a formal, symbolic model for it—using mathematical equations, programs like cellular automata, or whatever. We might think that that connection would immediately define an interpretation for our formal system. But once again it does not, because our model is just a model, that captures some features of the system, and idealizes others away. And seeing how that works again requires cultural context.

There is one slight exception to this: what if there is a fundamental theory of all of physics, that can perhaps be stated as a simple program? That program is then not just an idealized model, but a full representation of physics. And the point is that that "ground truth" about our universe describes the physics that govern absolutely any entity that exists in our universe.

If there is indeed a simple model for the universe, it's essentially inevitable that the things it directly describes are not ones familiar from our everyday sensory experience; for example they're presumably "below" constructs like space and time as we know them. But still, we might imagine that we could show off our achievements by presenting a version of the ultimate theory for our universe (if we'd found it!). But even with this, there's a problem. Because, well, it's not difficult to show a correct model for the universe: you just have to look at the actual universe! So the main information in an abstract representation is in what the primitives of the abstract representation end up being (do you set up your universe in terms of networks, or algebraic structures, or what?).

Let's back off from this level of philosophy for a moment. Let's say we're delivering a physical object—like a spacecraft, or a car—to our aliens. You might think the problem would be simpler. But the problem again is that it requires cultural context to decide what's important, and what's not. Is the placement of those rivets a message? Or an engineering optimization? Or an engineering tradition? Or just arbitrary?

Pretty much everything on, say, a spacecraft was presumably put there as part of building the spacecraft. Some was decided upon "on purpose" by its human designers. Some was probably a consequence of the physics of its manufacturing. But in the end the spacecraft just is what it is. You could imagine reconstructing the neural processes of its human designers, as you could imagine reconstructing the heat flows in the annealing of some part of it. But what is just the mechanism by which the spacecraft was built, and what is its "purpose"—or what is it trying to "communicate"?

The Molecular Version

It's one thing to talk about sending messages based on the achievements of our civilization. But what about just sending our DNA? Yes, it doesn't capture (at least in any direct way) all our intellectual achievements. But it does capture a couple of billion years of biological evolution, and represent a kind of memorial of the 10^{40} or so organisms that have ever lived on our planet.

Of course, we might again ask, "what does it mean?" And indeed one of the points of Darwinism is that the forms of organisms (and the DNA that defines them) arise purely as a consequence of the process of biological evolution, without any "intentional design." Needless to say, when we actually start talking about biological organisms there's a tremendous tendency to say things like "that mollusc has a pointy shell because it's useful in wedging itself in rocks"—in other words, to attribute a purpose to what has arisen from evolution.

So what would we be communicating by sending DNA (or, for that matter, complete instances of organisms)? In a sense we'd be providing a frozen representation of history, though now biological history. There's an issue of context again too. How does one interpret a disembodied piece of DNA? (Or, what environment is needed to get this spore to actually do something?)

Long ago it used to be said that if there were "organic molecules" out in space, it'd be a sign of life. But in fact plenty of even quite complex molecules have now been found, even in interstellar space. And while these molecules no doubt reflect all sorts of complex physical processes, nobody takes them as a sign of anything like life.

So what would happen if aliens found a DNA molecule? Is that elaborate sequence a "meaningful message," or just something created through random processes? Yes, in the end the sequences that have survived in modern DNA reflect in some way what leads to successful organisms in our specific terrestrial environment, though—just as with technology and language—there is a certain feedback in the way that organisms create the environment for others.

But, so, what does a DNA sequence show? Well, like a library of human knowledge, it's a representation of a lot of elaborate historical processes—and of a lot of irreducible computation. But the difference is that it doesn't have any "spark of human intention" in it.

Needless to say, as we've been discussing, it's hard to identify a signature for that. If we look at things we've created so far in our civilization, they're typically recognizable by the presence of things like (what we at least currently consider) simple geometrical forms, such as lines and circles and so on. And in a sense it's ironic that after all our development as a civilization, what we produce as artifacts look so much simpler than what nature routinely produces.

And we don't have to look at biology, with all its effort of biological evolution. We can just as well think of physics, and things like the forms of snowflakes or splashes or turbulent fluids.

As I've argued at length, the real point is that out in the computational universe of possible programs, it's actually easy to find examples where even simple underlying rules lead to highly complex behavior. And that's what's happening in nature. And the only reason we don't see that usually in the things we construct is that we constrain ourselves to use engineering practices that avoid complexity, so that we can foresee their outcome. And the result of this is that we tend to always end up with things that are simple and familiar.

Now that we understand more about the computational universe, we can see, however, that it doesn't always have to be this way. And in fact I have had great success just "mining the computational universe" for programs (and structures) that turn out to be useful, independent of whether one can "understand" how they operate. And something like the same thing happens when one trains a modern machine learning system. One ends up with a technological system that we can identify as achieving some overall purpose, but where the individual parts we can't particularly recognize as doing meaningful things.

And indeed my expectation is that in the future, a smaller and smaller fraction of human-created technology will be "recognizable" and "understandable." Optimized circuitry doesn't have nice repetitive structure; nor do optimized algorithms. Needless to say, it's sometimes hard to tell what's going on. Is that pattern of holes on a speakerphone arranged to optimize some acoustic feature, or is it just "decorative"?

Yet again we're thrust back into the same philosophical quandary: we can see the mechanism by which things operate, and we can come up with a story that describes why they might work that way. But there is no absolute way to decide whether that story is "correct"—except by referring back to the details of humans and human culture.

Talking about the World

Let's go back to language. What really is a language? Structurally (at least in all the examples we know so far) it's a collection of primitives (words, grammatical constructs, etc.) that can be assembled according to certain rules. And yes, we can look at a language formally at this level, just like we can look, say, at how to make tilings according to some set of rules. But what makes a language useful for communication is that its primitives somehow relate to the world—and that they're tied into knowledge.

In a first approximation, the words or other primitives in a language end up being things that are useful in describing aspects of the world that we want to communicate. We have different words for "table" and "chair" because those are buckets of meaning that we find it useful to distinguish. Yes, we could start describing the details of how the legs of the table are arranged, but for many purposes it's sufficient to just have that one word, or one symbolic primitive, "table", that describes what we think of as a table.

Of course, for the word "table" to be useful for communication, the sender and recipient of the word have to have shared understanding of its meaning. As a practical matter, for natural languages, this is usually achieved in an essentially societal way—with people seeing other people describing things as "tables".

How do we determine what words should exist? It's a societally driven process, but at some level it's about having ways to define concepts that are repeatedly useful to us. There's a certain circularity to the whole thing. The concepts that are useful to us depend on the environment in which we live. If there weren't any tables around (e.g. during the Stone Age), it wouldn't be terribly useful to have the word "table".

But then once we introduce a word for something (like "blog"), it starts to be easier for us to think about the thing—and then there tends to be more of it in the environment that we construct for ourselves, or choose to live in.

Imagine an intelligence that exists as a fluid (say the weather, for example). Or even imagine an aquatic organism, used to a fluid environment. Lots of the words we might take for granted about solid objects or locations won't be terribly useful. And instead there might be words for aspects of fluid flow (say, lumps of vorticity that change in some particular way) that we've never identified as concepts that we need words for.

It might seem as if different entities that exist within our physical universe must necessarily have some commonality in the way they describe the world. But I don't think this is the case—essentially as a consequence of the phenomenon of computational irreducibility.

The issue is that computational irreducibility implies that there are in effect an infinite number of irreducibly different environments that can be constructed on the basis of our physical universe—just like there are an infinite number of irreducibly different universal computers that can be built up using any given universal computer. In more practical terms, a way to say this is that different entities—or different intelligences—could operate using irreducibly different "technology stacks", based on different elements of the physical world (e.g. atomic vs. electronic vs. fluidic vs. gravitational, etc.) and different chains of inventions. And the result would be that their way of describing the world would be irreducibly different.

Forming a Language

But OK, given a certain experience of the world, how can one figure out what words or concepts are useful in describing it? In human natural languages, this seems to be something that basically just evolves through a process roughly analogous to natural selection in the course of societal use of the language. And in designing the Wolfram Language as a computational communication language I've basically piggybacked on what has evolved in human natural language.

So how can we see the emergence of words and concepts in a context further away from human language? Well, in modern times, there's an answer, which is basically to use our emerging example of alien intelligence: artificial intelligence.

Just take a neural network and start feeding it, say, images of lots of things in the world. (By picking the medium of 2D images, with a particular encoding of data, we're essentially defining ourselves to be "experiencing the world" in a specific way.) Now see what kinds of distinctions the neural net makes in clustering or classifying these images.

In practice, different runs will give different answers. But any pattern of answers is in effect providing an example of the primitives for a language.

An easy place to see this is in training an image identification network.* We started doing this several years ago with tens of millions of example images, in about 10,000 categories. And what's notable is that if you look inside the network, what it's effectively doing is to hone in on features of images that let it efficiently distinguish between different categories.

These features then in effect define the emergent symbolic language of the neural net. And, yes, this language is quite alien to us. It doesn't directly reflect human language or human thinking. It's in effect an alternate path for "understanding the world," different from the one that humans and human language have taken.

Can we decipher the language? Doing so would allow us to "explain the story" of what the neural net is "thinking." But it won't typically be easy to do. Because the "concepts" that are being identified in the neural network typically won't have easy translations to things we know about—and we'll be stuck in effect doing something like natural science to try to identify phenomena from which we can build up a description of what's going on.

OK, but in the problem of communicating with aliens, perhaps this suggests a way. Don't try (and it'll be hard) to specify a formal definition of "chair." Just show lots of examples of chairs—and use this to define the symbolic "chair" construct. Needless to say, as soon as one's showing pictures of chairs, not providing actual chairs, there are issues of how one's describing or encoding things. And while this approach might work decently for common nouns, it's more challenging for things like verbs, or more complex linguistic constructs.

But if we don't want our spacecraft full of sample objects (a kind of ontological Noah's Ark), maybe we could get away with just sending a device that looks at objects, and outputs what they're called. After all, a human version of this is basically how people learn languages, either as

---

* www.imageidentify.com

children, or when they're out doing linguistic fieldwork. And today we could certainly have a little computer with a very respectable, human-grade image identifier on it.

But here's the problem. The aliens will start showing the computer all sorts of things that they're familiar with. But there's no guarantee whatsoever that they'll be aligned with the things we (or the image identifier) have words for. One can already see the problem if one feeds an image identifier human abstract art; it's likely to be even worse with the products of alien civilization:

What the Wolfram Language Does

So can the Wolfram Language help? My goal in building it has been to create a bridge between the things humans want to do, and the things computation abstractly makes possible. And if I were building the language not for humans but for aliens—or even dolphins—I'd expect it to be different.

In the end, it's all about computation, and representing things computationally. But what one chooses to represent—and how one does it—depends on the whole context one's dealing with. And in fact, even for us humans, this has steadily changed over time. Over the 30+ years I've been working on the Wolfram Language, for example, both technology and the world have measurably evolved—with the result that there are all sorts of new things that make sense to have in the language. (The advance of our whole cultural understanding of computation—with things like hyperlinks and functional programming now becoming commonplace—also changes the concepts that can be used in the language.)

Right now most people think of the Wolfram Language mainly as a way for humans to communicate with computers. But I've always seen it as a general computational communication language for humans and computers—that's relevant among other things in giving us humans a way to think and communicate in computational terms. (And, yes, the kind of computational thinking this makes possible is going to be increasingly critical—even more so than mathematical thinking has been in the past.)

But the key point is that the Wolfram Language is capturing computation in human-compatible terms. And in fact we can view it as in effect giving a definition of which parts of the universe of possible computations we humans—at the current stage in the evolution of our civilization—actually care about.

Another way to put this is that we can think of the Wolfram Language as providing a compressed representation (or, in effect, a model) of the core content of our civilization. Some of that content is algorithmic and structural; some of it is data and knowledge about the details of our world and its history.

There's more to do to make the Wolfram Language into a full symbolic discourse language that can express a full range of human intentions (for example what's needed for encoding complete legal contracts, or ethical principles for AIs.) But with the Wolfram Language as it exists today, we're already capturing a very broad swath of the concerns and achievements of our civilization.

But how would we feed it to aliens? At some level its gigabytes of code and terabytes of data just define rules—like the rules for a cellular automaton or any other computational system. But the point is that these rules are chosen to be ones that do computations that we humans care about.

It's a bit like those Egyptian tomb models, which show things Egyptians cared about doing. If we give the aliens the Wolfram Language we're essentially giving them a computational model of things we care about doing. Except, of course, that by providing a whole language—rather than just individual pictures or dioramas—we're communicating in a vastly broader and deeper way.

The Reality of Time Capsules

What we're trying to create in a sense amounts to a time capsule. So what can we learn from time capsules of the past? Sadly, the history is not too inspiring.

Particularly following the discovery of King Tutankhamun's tomb in 1922, there was a burst of enthusiasm for time capsules that lasted a little over 50 years, and led to the creation—and typically burial—of perhaps 10,000 capsules. Realistically, though, the majority of these time capsules are even by now long forgotten—most often because the organizations that created them have changed or disappeared. (The Westinghouse Time Capsule for the 1939 World's Fair was at one time a proud example; but last year the remains of Westinghouse filed for bankruptcy.)

My own email archive records a variety of requests in earlier years for materials for time capsules, and looking at it today I'm reminded that we seem to have created a time capsule for Mathematica's 10th anniversary in 1998. But where is it now? I don't know. And this is a typical problem. Because whereas an ongoing archive (or library, etc.) can keep organized track of things, time capsules tend to be singular, and have a habit of ending up sequestered away in places that quickly get obscured and forgotten. (The reverse can also happen: people think there's a time capsule somewhere—like one supposedly left by John von Neumann to be opened 50 years after his death—but it turns out just to be a confusion.)

The one area where at least informal versions of time capsules seem to work out with some frequency is in building construction. In England, for example, when thatched roofs are redone after 50 years or so, it's common for messages from the previous workers to be found. But a particularly old tradition—dating even back to the Babylonians—is to put things in the foundations, and particularly at the cornerstones, of buildings.

Often in Babylonian times, there would just be an inscription cursing whoever had demolished the building to the point of seeing its foundations. But later, there was for example a longstanding tradition among Freemason stonemasons to embed small boxes of memorabilia in public buildings they built.

More successful, however, than cleverly hidden time capsules have been stone inscriptions out in plain sight. And indeed much of our knowledge of ancient human history and culture comes from just such objects. Sometimes they are part of large surviving architectural structures. But one famous example (key to the deciphering of cuneiform) is simply carved into the side of a cliff in what's now Iran:

For emphasis, it has a life-size relief of a bunch of warriors at the top. The translated text begins: "I am Darius the great, king of kings, …" and goes on to list 76 paragraphs of Darius's achievements, many of them being the putting down of attempted rebellions against him, in which he brought their leaders to sticky ends.

Such inscriptions were common in the ancient world (as their tamer successors are common today). But somehow their irony was well captured by my childhood favorite poem, Shelley's "Ozymandias" (named after Ramses II of Egypt):

> I met a traveller from an antique land,
> Who said—Two vast and trunkless legs of stone
> Stand in the desert.
>
> …
>
> And on the pedestal, these words appear:
> 'My name is Ozymandias, King of Kings;
> Look on my Works, ye Mighty, and despair!'
> Nothing beside remains. Round the decay
> Of that colossal Wreck, boundless and bare
> The lone and level sands stretch far away.

If there was a "Risks" section to a prospectus for the beacon project, this might be a good exhibit for it.

Of course, in addition to intentional "showoff" inscriptions, ancient civilizations left plenty of "documentary exhaust" that's still around in one form or another today. A decade ago, for example, I bought off the web (and, yes, I'm pretty sure it's genuine) a little cuneiform tablet from about 2100 BC:

It turns out to be a contract saying that a certain Mr. Lu-Nanna is receiving 1.5 gur (about 16 cubic feet) of barley in the month of Dumuzi (Tammuz/June–July), and that in return he should pay out certain goods in September–November.

Most surviving cuneiform tablets are about things like this. One in a thousand or so are about things like math and astronomy, though. And when we look at these tablets today, it's certainly interesting to see how far the Babylonians had got in math and astronomy. But (with the possible exception of some astronomical parameters) after a while we don't really learn anything more from such tablets.

And that's a lesson for our efforts now. If we put math or science facts in our beacons, then, yes, it shows how far we've got (and of course to make the best impression we should try to illustrate the furthest reaches of, for example, today's math, which will be quite hard to do). But it feels a bit like job applicants writing letters that start by explaining basic facts. Yes, we already know those; now tell us something about yourselves!

But what's the best way to do that? In the past the channel with the highest bandwidth was the written word. In today's world, maybe video—or AI simulation—goes further. But there's more—and we're starting to see this in modern archaeology. The fact is that pretty much any solid object carries microscopic traces of its history. Maybe it's a few stray molecules—say from the DNA of something that got onto an eating utensil. Maybe it's microscopic scratches or cracks in the material itself, indicating some pattern of wear.

Atomic force microscopy gives us the beginning of one way to systematically read such things out. But as molecular-scale computing comes online, such capabilities will grow rapidly. And this will give us access to a huge repository of "historical exhaust."

We won't immediately know the name "Lu-Nanna." But we might well know their DNA, the DNA of their scribe, what time of day their tablet was made, and what smells and maybe even sounds there were while the clay was drying. All of this one can think of as a form of "sensory data"— once again giving us information on "what happened," though with no interpretation of what was considered important.

Messages in Space

OK, but our objective is to put information about our civilization out into space. So what's the history of previous efforts to do that? Well, right now there are just four spacecraft outside our solar system (and another one that's headed there), and there are under 100 spacecraft more-or-less intact on various planetary surfaces (not counting hard landings, melted spacecraft on Venus, etc.). And at some level a spacecraft itself is a great big "message", illustrating lots of technology and so on.

Probably the largest amounts of "design information" will be in the microprocessors. And although radiation hardening forces deep space probes to use chip designs that are typically a decade or more behind the latest models, something like the *New Horizons* spacecraft launched in 2006 still has MIPS R3000 CPUs (albeit running at 12 MHz) with more than 100,000 transistors:

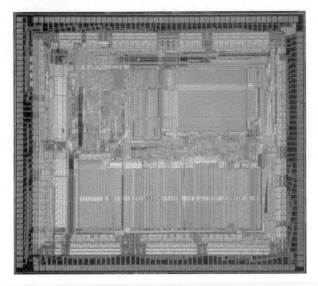

There are also substantial amounts of software, typically stored in some kind of ROM. Of course, it may not be easy to understand, even for humans—and indeed just last month, firing backup thrusters on *Voyager 1* that hadn't been used for 37 years required deciphering the machine code for a long-extinct custom CPU.

The structure of a spacecraft tells a lot about human engineering and its history. Why was the antenna assembly that shape? Well, because it came from a long lineage of other antennas that were conveniently modeled and manufactured in such-and-such a way, and so on.

But what about more direct human information? Well, there are often little labels printed on components by manufacturers. And in recent times there's been a trend of sending lists of people's names (more than 400,000 on *New Horizons*) in engravings, microfilm, or CDs/DVDs. (The MAVEN Mars mission also notably carried 1000+ publicly submitted haikus about Mars, together with 300+ drawings by kids, all on a DVD.) But on most spacecraft the single most prominent piece of "human communication" is a flag:

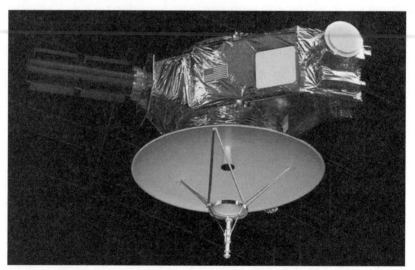

A few times, however, there have been explicit, purposeful plaques and things displayed. For example, on the leg of *Apollo 11*'s lunar module this was attached (with the Earth rendered in a stereographic projection cut in the middle of the Atlantic around 20°W):

Each *Apollo* mission to the Moon also planted an American flag (most still "flying" according to recent high-res reconnaissance)—strangely reminiscent of shrines to ancient gods found in archaeological remains:

The very first successful moon probe (*Luna 2*) carried to the Moon this ball-like object—which was intended to detonate like a grenade and scatter its pentagonal facets just before the probe hit the lunar surface, proclaiming (presumably to stake a claim): "USSR, January 1959":

On Mars, there's a plaque that seems more like the cover sheet for a document—or that might be summarized as "putting the output of some human cerebellums out in the cosmos" (what kind of personality analysis could the aliens do from those signatures?):

There's another list of names, this time an explicit memorial for fallen astronauts, left on the Moon by *Apollo 15*. But this time it comes with a small figurine, strangely reminiscent of the figurines we find in early archaeological remains:

Figurines have actually been sent on other spacecraft too. Here are some LEGO ones that went to Jupiter on the *Juno* spacecraft (from left to right: mythological Jupiter, mythological Juno, and real Galileo, complete with LEGO attachments):

Also on that spacecraft was a tribute to Galileo—though all this will be vaporized when the spacecraft deorbits Jupiter in a few years to avoid contaminating any moons:

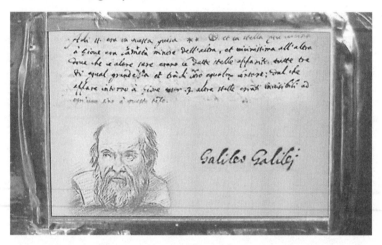

A variety of somewhat random personal and other trinkets have been left—usually unofficially—on the Moon. An example is a collection of tiny artworks (which are head scratchers even for me as a human) apparently attached to the leg of the *Apollo 12* lunar module:

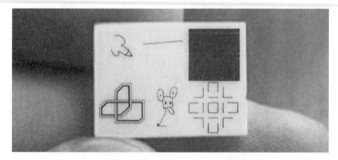

There was also a piece of "artwork" (doubling as a color calibration target) sent on the ill-fated *Beagle 2* Mars lander:

There are "MarsDials" on several Mars landers, serving as sundials and color calibration targets. The earlier ones had the statement "Two worlds, one sun"—along with the word "Mars" in 22 languages; on later ones the statement was the less poetic "On Mars, to explore":

As another space trinket, the *New Horizons* spacecraft that recently passed Pluto has a simple Florida state quarter on board—which at least was presumably easy and cheap to obtain near its launch site.

But the most serious—and best-known—attempts to provide messages are the engraved aluminum plaques on the *Pioneer 10* and *11* spacecraft that were launched in 1972 and 1973 (though are sadly now out of contact):

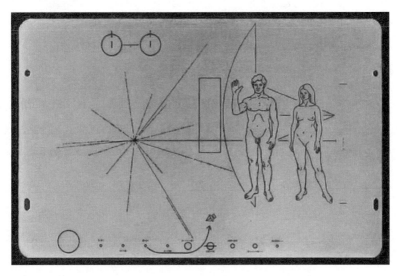

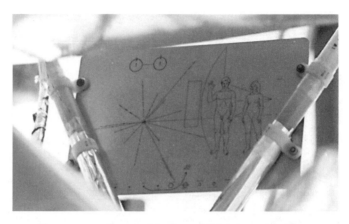

I must say I have never been a big fan of this plaque. It always seemed to me too clever by half. My biggest beef has always been with the element at the top left. The original paper (with lead author Carl Sagan) about the plaque states that this "should be readily recognizable to the physicists of other civilizations."

But what is it? As a human physicist, I can figure it out: it's an iconic representation of the hyperfine transition of atomic hydrogen—the so-called 21-centimeter line. And those little arrows are supposed to represent the spin directions of protons and electrons before and after the transition. But wait a minute: electrons and protons are spin-1/2, so they act as spinors. And yes, traditional human quantum mechanics textbooks do often illustrate spinors using vectors. But that's a really arbitrary convention.

Oh, and why should we represent quantum mechanical wavefunctions in atoms using localized lines? Presumably the electron is supposed to "go all the way around" the circle, indicating that it's delocalized. And, yes, you can explain that iconography to someone who's used to human quantum mechanics textbooks. But it's about as obscure and human-specific as one can imagine. And, by the way, if one wants to represent 21.106-centimeter radiation, why not just draw a line precisely that length, or make the plaque that size (it actually has a width of 22.9 centimeters)!

I could go on and on about what's wrong with the plaque. The rendering conventions for the (widely mocked) human figures, especially when compared to those for the spacecraft. The use of an arrow to show the spacecraft direction (do all aliens go through a stage of shooting arrowheads?). The trailing (binary) zeros to cover the lack of precision in pulsar periods.

The official key from the original paper doesn't help the case, and in fact the paper lays out some remarkably elaborate "science IQ test" reasoning needed to decode other things on the plaque:

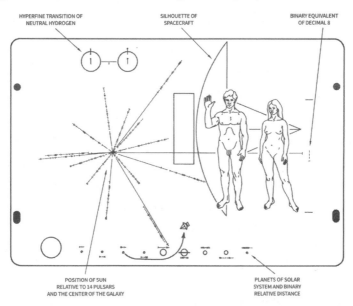

After the attention garnered by the *Pioneer* plaques, a more ambitious effort was made for the *Voyager* spacecraft launched in 1977. The result was the 12-inch gold-plated Voyager Golden Record, with an "album cover":

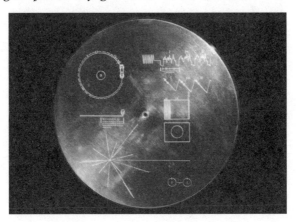

In 1977, phonograph records seemed like "universally obvious technology." Today of course even the concept of analog recording is (at least for now) all but gone. And what of the elaborately drawn "needle" on the top left? In modern times the obvious way to read the record would just be to image the whole thing, without any needles tracking grooves.

But, OK, so what's on the record? There are some spoken greetings in 55 languages (beginning with one in a modern rendering of Akkadian), along with a 90-minute collection of music from around the world. (Somehow I imagine an alien translator—or, for that matter, an AI— trying in vain to align the messages between the words and the music.) There's an hour of recorded brainwaves of Carl Sagan's future wife (Ann Druyan), apparently thinking about various things.

Then there are 116 images, encoded in analog scan lines (though I don't know how color was done). Many were photographs of 1970s life on Earth. Some were "scientific explanations," which are at least good exercises for human science students of the 2010s to interpret (though the real-number rounding is weird, there are "9 planets," there's "S" in place of "C" as a base pair—and it's charming to see the stencil-and-ink rendering):

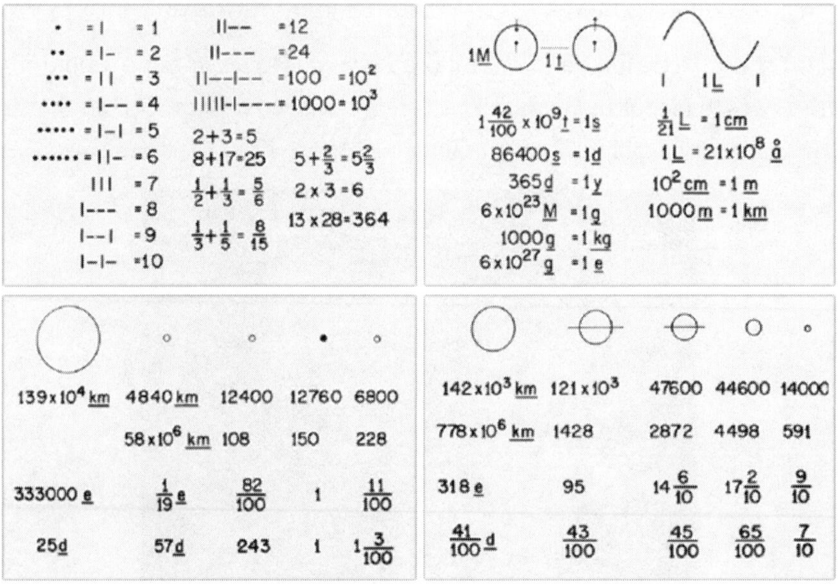

Panel 1 (numerals and arithmetic):

$$\bullet = | = 1 \qquad ||-- = 12$$
$$\bullet\bullet = |- = 2 \qquad ||--- = 24$$
$$\bullet\bullet\bullet = || = 3 \qquad ||--|-- = 100 = 10^2$$
$$\bullet\bullet\bullet\bullet = |-- = 4 \qquad |||||-|---- = 1000 = 10^3$$
$$\bullet\bullet\bullet\bullet\bullet = |-| = 5$$
$$\bullet\bullet\bullet\bullet\bullet\bullet = ||- = 6 \qquad 2+3=5$$
$$||| = 7 \qquad 8+17=25 \qquad 5+\tfrac{2}{3}=5\tfrac{2}{3}$$
$$|--- = 8 \qquad \tfrac{1}{2}+\tfrac{1}{3}=\tfrac{5}{6} \qquad 2\times3=6$$
$$|--| = 9 \qquad \tfrac{1}{3}+\tfrac{1}{5}=\tfrac{8}{15} \qquad 13\times28=364$$
$$|-|- = 10$$

Panel 2 (units):

$$1M \quad 11 \qquad 1L$$
$$1\tfrac{42}{100}\times10^9\,t = 1s \qquad \tfrac{1}{21}L = 1cm$$
$$86400s = 1d \qquad 1L = 21\times10^8\,\mathring{a}$$
$$365d = 1y \qquad 10^2\,cm = 1m$$
$$6\times10^{23}\,M = 1g \qquad 1000m = 1km$$
$$1000g = 1kg$$
$$6\times10^{27}\,g = 1e$$

Panel 3 (planets):

139×10^4 km	4840 km	12400	12760	6800
	58×10^6 km	108	150	228
333000 e	$\tfrac{1}{19}$ e	$\tfrac{82}{100}$	1	$\tfrac{11}{100}$
25d	57d	243	1	$1\tfrac{3}{100}$

Panel 4 (planets):

142×10^3 km	121×10^3	47600	44600	14000
778×10^6 km	1428	2872	4498	591
318 e	95	$14\tfrac{6}{10}$	$17\tfrac{2}{10}$	$\tfrac{9}{10}$
$\tfrac{41}{100}$ d	$\tfrac{43}{100}$	$\tfrac{45}{100}$	$\tfrac{65}{100}$	$\tfrac{7}{10}$

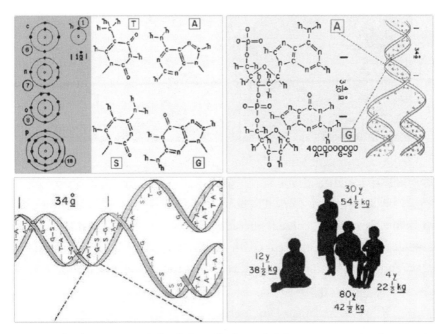

Yes, when I proposed the "alien flashcards" for scientists in the movie *Arrival*, I too started with binary—though in modern times it's easy and natural to show the whole nested pattern of successive digit sequences:

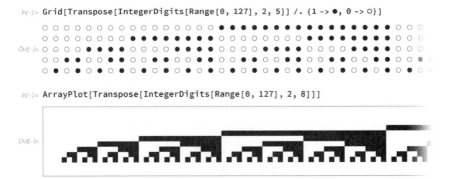

Among efforts after *Voyager* have been the (very 1990s-style) CD of human Mars-related "Visions of Mars" fiction on the failed 1996 *Mars 96* spacecraft, as well as the 2012 "time capsule" CD of images and videos on the EchoStar 16 satellite in geostationary orbit around Earth:

A slightly different kind of plaque was launched back in 1976 on the LAGEOS-1 satellite that's supposed to be in polar orbit around the Earth for 8.4 million years. There are the binary numbers, reminiscent of Leibniz's original "binary medal." And then there's an image of the predicted effect of continental drift (and what about sea level?) from 2^{28} years ago, to the end of the satellite's life—that to me gives off a certain "so, did we get it right?" vibe:

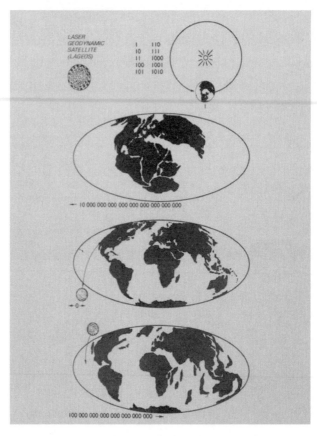

There was almost an engraved diamond plaque sent on the Cassini mission to Saturn and beyond in 1997, but as a result of human disagreements, it was never sent—and instead, in a very Ozymandias kind of way, all that's left on the spacecraft is an empty mounting pedestal, whose purpose might be difficult to imagine.

Still another class of artifacts sent into the cosmos are radio transmissions. And until we have better-directed radio communications (and 5G will help), we're radiating a certain amount of (increasingly encrypted) radio energy into the cosmos. The most intense ongoing transmissions remain the 50 Hz or 60 Hz hum of power lines, as well as the perhaps-almost-pulsar-like Ballistic Missile Early Warning System radars. But in the past there've been specific attempts to send messages for aliens to pick up.

The most famous was sent by the Arecibo radio telescope in 1974. Its repetition length was a product of two primes, intended to suggest assembly as a rectangular array. It's an interesting exercise for humans to try to decipher the resulting image. Can you see the sequence of binary numbers? The schematic DNA, and the bitvectors for its components? The telescope icon? And the little 8-bit-video-game-like human?

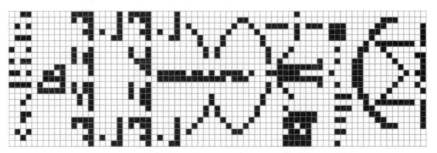

There've been other messages sent, including a Doritos ad, a Beatles song, some Craigslist pages, and a plant gene sequence—as well as some arguably downright embarrassing "artworks.")

Needless to say, we pick up radio transmissions from the cosmos that we don't understand fairly often. But are they signs of intelligence? Or "merely physics"? As I've said, the Principle of Computational Equivalence tells us there isn't ultimately a distinction. And that, of course, is the challenge of our beacons project.

It's worth mentioning that in addition to what's been sent into space, there are a few messages on Earth specifically intended for at least few thousand years in the future. Examples are the 2000-year equinox star charts at the Hoover Dam, and the long-planned-but-not-yet-executed 10,000-year "stay away; it's radioactive" warnings (or maybe it's an "atomic priesthood" passing information generation to generation) for facilities like the WIPP nuclear waste repository in southeastern New Mexico. (Not strictly a "message", but there's also the "10,000-year clock" being built in West Texas.)

A discussion of extraterrestrial communication wouldn't be complete without at least mentioning the 1960 book *Lincos: Design of a Language for Cosmic Intercourse*—my copy of which wound up on the set of *Arrival*. The idea of the book was to use the methods and notation of mathematical logic to explain math, science, human behavior, and other things "from first principles." Its author, Hans Freudenthal, had spent decades working on math education—and on finding the best ways to explain math to (human) kids.

Lincos was created too early to benefit from modern thinking about computer languages. And as it was, it used the often almost comically abstruse approach of Whitehead and Russell's 1910 *Principia Mathematica*—in which even simple ideas become notationally complex. When it came to a topic like human behavior Lincos basically just gave examples, like small scenes in a stage play—but written in the notation of mathematical logic.

Yes, it's interesting to try to have a symbolic representation for such things—and that's the point of my symbolic discourse language project. But even though Lincos was at best just at the very beginning of trying to formulate something like this, it was still the obvious source for attempts to send "active SETI" messages starting in 1999, and some low-res bitmaps of Lincos were transmitted to nearby stars.

Science Fiction and Beyond

For our beacons project, we want to create human artifacts that will be recognized even by aliens. The related question of how alien artifacts might be recognizable has been tackled many times in science fiction.

Most often there's something that just "doesn't look natural," either because it's obviously defying gravity, or because it's just too simple or perfect. For example, in the movie *2001*, when the black cuboid monolith with its exact 1:4:9 side ratios shows up on Stone Age Earth or on the Moon, it's obvious it's "not natural."

On the flip side, people in the 1800s argued that the fact that, while complex, a human-made pocket watch was so much simpler than a biological organism meant that the latter could only be an "artifact of God." But actually I think the issue is just that our technology isn't advanced enough yet. We're still largely relying on engineering traditions and structures where we readily foresee every aspect of how our system will behave.

But I don't think this will go on much longer. As I've spent many years studying, out in the computational universe of all possible programs it's very common that the most efficient programs for a particular purpose don't look at all simple in their behavior (and in fact this is a somewhat inevitable consequence of making better use of computational resources). And the result is that as soon as we can systematically mine such programs (as Darwinian evolution and neural network training already begin to), we'll end up with artifacts that no longer look simple.

Ironically—but not surprisingly, given the Principle of Computational Equivalence—this suggests that our future artifacts will often look much more like "natural systems." And indeed our current artifacts may look as primitive in the future as many of those produced before modern manufacturing look to us today.

Some science fiction stories have explored "natural-looking" alien artifacts, and how one might detect them. Of course it's mired in the same issues that I've been exploring throughout this chapter—making it very difficult for example to tell for certain even whether the strangely red and strangely elongated interstellar object recently observed crossing our solar system is an alien artifact, or just a "natural rock."

The Space of All Possible Civilizations

A major theme of this chapter has been that "communication" requires a certain sharing of "cultural context." But how much sharing is enough? Different people—with at least fairly different backgrounds and experiences—can usually understand each other well enough for society to function, although as the "cultural distance" increases, such understanding becomes more and more difficult.

Over the course of human history, one can imagine a whole net of cultural contexts, defined in large part (at least until recently) by place and time. Neighboring contexts are typically closely connected—but to get a substantial distance, say in time, often requires following a quite long chain of intermediate connections, a bit like one might have to go through a chain of intermediate translations to get from one language to another.

Particularly in modern times, cultural context often evolves quite significantly even over the course of a single human lifetime. But usually the process is gradual enough that an individual can bridge the contexts they encounter—though of course there's no lack of older people who are at best confused at the preferences and interests of the young (think modern social media, etc.). And indeed were one just suddenly to wake up a century hence, it's fairly certain that some of the cultural context would be somewhat disorientingly different.

But, OK, can we imagine making some kind of formal theory of cultural contexts? To do so would likely in effect require describing the space of all possible civilizations. And at first this might seem utterly infeasible.

But when we explore the computational universe of possible programs we are looking at a space of all possible rules. And it's easy to imagine defining at least some feature of a civilization by some appropriate rule—and different rules can lead to dramatically different behavior, as in these cellular automata:

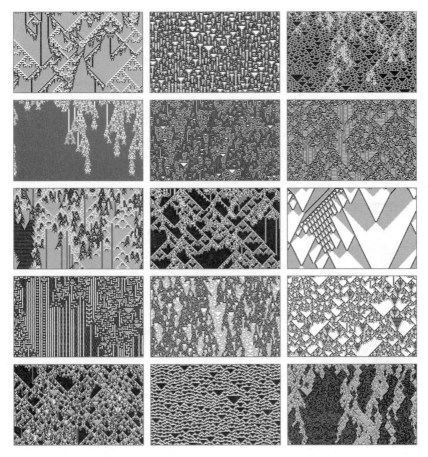

But, OK, what would "communication" mean in this context? Well, as soon as these rules are computationally universal (and the Principle of Computational Equivalence implies that except in trivial cases they always will be), there's got to be some way to translate between them. More specifically, given one universal rule, there must be some program for it—or some class of initial conditions—that make it emulate any other specified rule. Or, in other words, it must be possible to implement an interpreter for any given rule in the original rule.

We might then think of defining a distance between rules to be determined by the size or complexity of the interpreter necessary to translate between them. But while this sounds good in principle, it's certainly not an easy thing to deal with out in practice. And it doesn't help that interpretability can be formally undecidable, so there's no upper bound on the size or complexity of the translator between rules.

But at least conceptually, this gives us a chance to think about how a "communication distance" might be defined. And perhaps one could imagine a first approximation for the simplified case of neural networks, in which one just asks how difficult it is to train one network to act like another.

As a more down-to-earth analogy to the space of cultural contexts, we could consider human languages, of which there are about 10,000 known. One can assess similarities between languages by looking at their words, and perhaps by looking at things like their grammatical structures. And even though in first approximation all languages can talk about the same kinds of things, languages can at least superficially have significant differences.

But for the specific case of human languages, there's a lot determined by history. And indeed there's a whole evolutionary tree of languages that one can identify, that effectively explains what's close and what's not. (Languages are often related to cultures, but aren't the same. For example, Finnish is very different as a language from Swedish, even though Finnish and Swedish cultures are fairly similar.)

In the case of human civilizations, there are all sorts of indicators of similarity one might use. How similar do their artifacts look, say as recognized by neural networks? How similar are their social, economic, or genealogical networks? How similar are quantitative measures of their patterns of laws or government?

Of course, all human civilizations share all sorts of common history—and no doubt occupy only some infinitesimal corner in the space of all possible civilizations. And in the vast majority of potential alien civilizations, it's completely unrealistic to expect that the kinds of indicators we're discussing for human civilizations could even be defined.

So how might one characterize a civilization and its cultural context? One way is to ask how it uses the computational universe of possible programs. What parts of that universe does it care about, and what not?

Now perhaps the endpoint of cultural evolution is to make use of the whole space of possible programs. Of course, our actual physical universe is presumably based on specific programs—although within the universe one can perfectly well emulate other programs.

And presumably anything that we could identify as a definite "civilization" with definite "culture context" must make use of some particular type of encoding—and in effect some particular type of language—for the programs it wants to specify. So one way to characterize a civilization is to imagine what analog of the Wolfram Language (or in general what symbolic discourse language) it would invent to describe things.

Yes, I've spent much of my life building the single example of the Wolfram Language intended for humans. And now what I'm suggesting is to imagine the space of all possible analogous languages, with all possible ways of sampling and encoding the computational universe.

But that's the kind of thing we need to consider if we're serious about alien communication. And in a sense just as we might say that we're only going to consider aliens who live within a certain number of light years of us, so also we may have to say that we'll only consider aliens where the language defining their cultural context is within a certain "translation distance" of ours.

How can we study this in practice? Well, of course we could think about what analog of the Wolfram Language other creatures with whom we share the Earth might find useful. We could also think about what AIs would find useful—though there is some circularity to this, insofar as we are creating AIs for the purpose of furthering our human goals. But probably the best path forward is just to imagine some kind of abstract enumeration of possible Wolfram-Language analogs, and then to start studying what methods of translation might be possible between them.

What Should We Actually Send?

OK, so there are lots of complicated intellectual and philosophical issues. But if we're going to send beacons about the achievements of our civilization into space, what's the best thing to do in practice?

A few points are obvious. First, even though it might seem more "universal," don't send lots of content that's somehow formally derivable. Yes, we could say 2+2=4, or state a bunch of mathematical theorems, or show the evolution of a cellular automaton. But other than demonstrating that we can successfully do computation (which isn't anything special, given the Principle of Computational Equivalence) we're not really communicating anything like this. In fact, the only real information about us is our choice of what to send: which arithmetic facts, which theorems, etc.

Here's an ancient Egyptian die. And, yes, it's interesting that they knew about icosahedra, and chose to use them. But the details of the icosahedral shape don't tell us anything: it's just the same as any other icosahedron.

OK, so an important principle is: if we want to communicate about ourselves, send things that are special to us—which means all sorts of arbitrary details about our history and interests. We could send an encyclopedia. Or if we have more space, we could send the whole content of the web, or scans of all books, or all available videos.

There's a point, though, at which we will have sent enough: where basically there's the raw material to answer any reasonable question one could ask about our civilization and our achievements.

But how does one make this as efficient as possible? Well, at least for general knowledge I've spent a long time trying to solve that problem. Because in a sense that's what Wolfram|Alpha is all about: creating a system that can compute the answers to as broad a range as possible of questions.

So, yes, if we send a Wolfram|Alpha, we're sending knowledge of our civilization in a concentrated, computational form, ready to be used as broadly as possible.

Of course, at least the public version of Wolfram|Alpha is just about general, public knowledge. So what about more detailed information about humans and the human condition?

Well, there're always things like email archives, and personal analytics, and recordings, and so on. And, yes, I happen to have three decades of rather extensive data about myself, that I've collected mostly because it was easy for me to do.

But what could one get from that? Well, I suspect there's enough data there that at least in principle one could construct a bot of me from it: in other words, one could create an AI system that would respond to things in pretty much the same way I would.

Of course, one could imagine just "going to the source" and starting to read out the content of a human brain. We don't know how to do that yet. But if we're going to assume that the recipients of our beacons have advanced further, then we have to assume that given a brain, they could tell what it would do.

Indeed, perhaps the most obvious thing to send (though it's a bit macabre) would just be whole cryonically preserved humans (and, yes, they should keep well at the temperature of interstellar space!). Of course, it's ironic how similar this is to the Egyptian idea of making mummies—though our technology is better (even if we still haven't yet solved the problem of cryonics).

Is there a way to do even better, though? Perhaps by using AI and digital technology, rather than biology. Well, then we have a different problem. Yes, I expect we'll be able to make AIs that represent any aspect of our civilization that we want. But then we have to decide what the "best of our civilization" is supposed to be.

It's very related to questions about the ethics and "constitution" we should define for the AIs—and it's an issue that comes back directly to the dynamics of our society. If we were sending biological humans then we'd get whatever bundle of traits each human we sent happened to have. But if we're sending AIs, then somehow we'd have to decide which of the infinite range of possible characteristics we'd assign to best represent our civilization.

Whatever we might send—biological or digital—there's absolutely no guarantee of any successful communication. Sure, our person or our AI might do their best to understand and respond to the alien that picked them up. But it might be hopeless. Yes, our representative might be able to identify the aliens, and observe the computations they're doing. But that doesn't mean that there's enough alignment to be able to communicate anything we might think of as meaning.

It's certainly not encouraging that we haven't yet been able to recognize what we consider to be signs of extraterrestrial intelligence anywhere else in the universe. And it's also not encouraging that even on our own planet we haven't succeeded in serious communication with other species.

But just like Darius—or even Ozymandias—we shouldn't give up. We should think of the beacons we send as monuments. Perhaps they will be useful for some kind of "afterlife." But for now they serve as a useful rallying point for thinking about what we're proud of in the achievements of our civilization—and what we want to capture and celebrate in the best way we can. And I'll certainly be pleased to contribute to this effort the computational knowledge that I've been responsible for accumulating.

Pi or Pie?! Celebrating Pi Day of the Century

March 12, 2015

This coming Saturday is "Pi Day of the Century." The date 3/14/15 in month/day/year format is like the first digits of π=3.1415.... And at 9:26:53.589... it's a "super pi moment."

Between Mathematica and Wolfram|Alpha, I'm pretty sure our company has delivered more π to the world than any other organization in history. So of course we have to do something special for Pi Day of the Century.

A Corporate Confusion

One of my main roles as CEO is to come up with ideas—and I've spent decades building an organization that's good at turning those ideas into reality. Well, a number of weeks ago I was in a meeting about upcoming corporate events, and someone noted that Pi Day (3/14) would happen during the big annual SXSW (South by Southwest) event in Austin, Texas. So I said (or at least I thought I said), "We should have a big pi to celebrate Pi Day."

I didn't give it another thought, but a couple of weeks later we had another meeting about upcoming events. One agenda item was Pi Day. And the person who runs our Events group started talking about the difficulty of finding a bakery in Austin to make something suitably big. "What are you talking about?" I asked. And then I realized: "You've got the wrong kind of pi!"

I guess in our world pi confusions are strangely common. Siri's voice-to-text system sends Wolfram|Alpha lots of "pie" every day that we have to specially interpret as "pi." And then there's the Raspberry Pi, that has the Wolfram Language included. And for me there's the additional confusion that my personal fileserver happens to have been named "pi" for many years.

After the pi(e) mistake in our meeting we came up with all kinds of wild ideas to celebrate Pi Day. We'd already rented a small park in the area of SXSW, and we wanted to make the most interesting "pi countdown" we could. We resolved to get a large number of edible pie "pixels," and use them to create a π shape inside a pie shape. Of course, there'll be the obligatory pi selfie station, with a "Stonehenge" pi. And a pi(e)-decorated Wolfie mascot for additional selfies. And of course we'll be doing things with Raspberry Pis too.

A Piece of Pi for Everyone

I'm sure we'll have plenty of good "pi fun" at SXSW. But we also want to provide pi fun for other people around the world. We were wondering, "What can one do with pi?" Well, in some sense, you can do anything with pi. Because, apart from being the digits of pi, its infinite digit sequence is—so far as we can tell—completely random. So for example any run of digits will eventually appear in it.

How about giving people a personal connection to that piece of math? Pi Day is about a date that appears as the first digits of pi. But any date appears somewhere in pi. So, we thought: Why not give people a way to find out where their birthday (or other significant date) appears in pi, and use that to make personalized pi T-shirts and posters?

In the Wolfram Language, it's easy to find out where your birthday appears in π. It's pretty certain that any mm/dd/yy will appear somewhere in the first 10 million digits. On my desktop computer (a Mac Pro), it takes 6.28 seconds (2π?!) to compute that many digits of π.

Here's the Wolfram Language code to get the result and turn it into a string (dropping the decimal point at position 2):

```
In[ ]:= PiString = StringDrop[ToString[N[Pi, 10^7]], {2}];
```

Now it's easy to find any "birthday string":

```
In[ ]:= First[StringPosition[PiString, "82959"]]
Out[ ]= {151653, 151657}
```

So, for example, my birthday string first appears in π starting at digit position 151,653.

What's a good way to display this? It depends how "pi lucky" you are. For those born on 4/15/92, their birthdate already appears at position 3. (Only about a certain fraction of positions correspond to a possible date string.) People born on November 23, 1960, have the birthday string that's farthest out, appearing only at position 9,982,546. And in fact most people have birthdays that are pretty "far out" in π (the average is 306,150 positions).

Our longtime art director had the idea of using a spiral that goes in and out to display the beginnings and ends of such long digit sequences. And almost immediately, he'd written the code to do this (one of the great things about the Wolfram Language is that non-engineers can write their own code...).

Next came deploying that code to a website. And thanks to the Wolfram Cloud, this was basically just one line of code! So now you can go to MyPiDay.com...

... and get your own piece of π!

The Science of Pi

With all this discussion about pi, I can't resist saying just a little about the science of pi. But first, just why is pi so famous? Yes, it's the ratio of circumference to diameter of a circle. And that means that π appears in zillions of scientific formulas. But it's not the whole story. (And for example most people have never even heard of the analog of π for an ellipse—a so-called complete elliptic integral of the second kind.)

The bigger story is that π appears in a remarkable range of mathematical settings—including many that don't seem to have anything to do with circles. Like sums of negative powers, or limits of iterations, or the probability that a randomly chosen fraction will not be in lowest terms.

If one's just looking at digit sequences, pi's 3.1415926... doesn't seem like anything special. But let's say one just starts constructing formulas at random and then doing traditional mathematical operations on them, like summing series, doing integrals, finding limits, and so on. One will get lots of answers that are 0, or 1/2, or $\sqrt{2}$. And there'll be plenty of cases where there's no closed form one can find at all. But when one can get a definite result, my experience is that it's remarkably common to find π in it.

A few other constants show up too, like e (2.1718...), or Euler gamma (0.5772...), or Catalan's constant (0.9159...). But π is distinctly more common.

Perhaps math could have been set up differently. But at least with math as we humans have constructed it, the number that is π is a widespread building block, and it's natural that we gave it a name, and that it's famous—now even to the point of having a day to celebrate it.

What about other constants? "Birthday strings" will certainly appear at different places in different constants. And just like when Wolfram|Alpha tries to find closed forms for numbers, there's typically a tradeoff between digit position and obscurity of the constants used. So, for example, my birthday string appears at position 151,653 in π, 241,683 in e, 45,515 in $\sqrt{2}$, 40,979 in $\zeta(3)$ ..., and 196 in the 1601th Fibonacci number.

Randomness in π

Let's say you make a plot that goes up whenever a digit of π is 5 or above, and down otherwise:

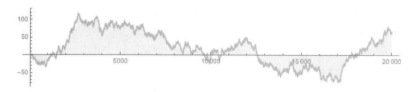

It looks just like a random walk. And in fact, all statistical and cryptographic tests of randomness that have been tried on the digits (except tests that effectively just ask "are these the digits of pi?") say that they look random too.

Why does that happen? There are fairly simple procedures that generate digits of pi. But the remarkable thing is that even though these procedures are simple, the output they produce is complicated enough to seem completely random. In the past, there wasn't really a context for thinking about this kind of behavior. But it's exactly what I've spent many years studying in all kinds of systems—and wrote about in *A New Kind of Science*. And in a sense the fact that one can "find any birthday in pi" is directly connected to concepts like my general Principle of Computational Equivalence.

SETI among the Digits

Of course, just because we've never seen any regularity in the digits of pi, it doesn't mean that no such regularity exists. And in fact it could still be that if we did a big search, we might find somewhere far out in the digits of pi some strange regularity lurking.

What would it mean? There's a science fiction answer at the end of Carl Sagan's book version of *Contact*. In the book, the search for extraterrestrial intelligence succeeds in making contact with an interstellar civilization that has created some amazing artifacts—and that then explains that what they in turn find remarkable is that encoded in the distant digits of pi, they've found intelligent messages, like an encoded picture of a circle.

At first one might think that finding "intelligence" in the digits of pi is absurd. After all, there's just a definite simple algorithm that generates these digits. But at least if my suspicions are correct, exactly the same is actually true of our whole universe, so that every detail of its history is in principle computable much like the digits of pi.

Now we know that within our universe we have ourselves as an example of intelligence. SETI is about trying to find other examples. The goal is fairly well defined when the search is for "human-like intelligence." But—as my Principle of Computational Equivalence suggests—I think that beyond that it's essentially impossible to make a sharp distinction between what should be considered "intelligent" and what is "merely computational."

If the century-old mathematical suspicion is correct that the digits of pi are "normal," it means that every possible sequence eventually occurs among the digits, including all the works of Shakespeare, or any other artifact of any possible civilization. But could there be some other structure—perhaps even superimposed on normality—that for example shows evidence of the generation of intelligence-like complexity?

While it may be conceptually simple, it's certainly more bizarre to contemplate the possibility of a human-like intelligent civilization lurking in the digits of pi, than in the physical universe as explored by SETI. But if one generalizes what one counts as intelligence, the situation is a lot less clear.

Of course, if we see a complex signal from a pulsar magnetosphere we say it's "just physics," not the result of the evolution of a "magnetohydrodynamic civilization." And similarly if we see some complex structure in the digits of pi, we're likely to say it's "just mathematics," not the result of some "number theoretic civilization."

One can generalize from the digit sequence of pi to representations of any mathematical constant that is easy to specify with traditional mathematical operations. Sometimes there are simple regularities in those representations. But often there is apparent randomness. And the project of searching for structure is quite analogous to SETI in the physical universe. (One difference, however, is that π as a number to study is selected as a result of the structure of our physical universe, our brains, and our mathematical development. The universe presumably has no such selection, save implicitly from the fact that we exist in it.)

I've done a certain amount of searching for regularities in representations of numbers like π. I've never found anything significant. But there's nothing to say that any regularities have to be at all easy to find. And there's certainly a possibility that it could take a SETI-like effort to reveal them.

But for now, let's celebrate the Pi Day of our century, and have fun doing things like finding birthday strings in the digits of pi. Of course, someone like me can't help but wonder what success there will have been by the next Pi Day of the Century, in 2115, in either SETI or "SETI among the digits"....

What Is Ultimately Possible in Physics?

October 9, 2009

The history of technology is littered with examples of things that were claimed to be impossible—but later done. So what is genuinely impossible in physics? There is much that we will not know about the answer to this question until we know the ultimate theory of physics. And even when we do—assuming it is possible to find it—it may still often not be possible to know what is possible.

Let's start, though, with the simpler question of what is possible in mathematics.

In the history of mathematics, particularly in the 1800s, many "impossibility results" were found. Squaring the circle. Trisecting an angle. Solving a quintic equation. But these were not genuine impossibilities. Instead, they were in a sense only impossibilities at a certain level of mathematical technology.

It is true, for example, that it is impossible to solve any quintic—if one is only allowed to use square roots and other radicals. But it is perfectly possible to write down a finite formula for the solution to any quintic in terms, say, of elliptic functions. And indeed, by the early 1900s, there emerged the view that there would ultimately be no such impossibilities in mathematics. And that instead it would be possible to build more and more sophisticated formal structures that would eventually allow any imaginable mathematical operation to be done in some finite way.

Yes, one might want to deal with infinite series or infinite sets. But somehow these could be represented symbolically, and everything about them could be worked out in some finite way.

In 1931, however, it became clear that this was not correct. For Gödel's theorem showed that in a sense mathematics can never be reduced to a finite activity. Starting from the standard axiom system for arithmetic and basic number theory, Gödel's theorem showed that there are questions that cannot be guaranteed to be answered by any finite sequence of mathematical steps—and that are therefore "undecidable" with the axiom system given.

One might still have thought that the problem was in a sense one of "technology": that one just needed stronger axioms, and then everything would be possible. But Gödel's theorem showed that no finite set of axioms can ever be added to cover all possible questions within standard mathematical theories.

At first, it wasn't clear how general this result really was. There was a thought that perhaps something like a transfinite sequence of theories could exist that would render everything possible—and that perhaps this might even be how human minds work.

But then in 1936 along came the Turing machine, and with it a new understanding of possibility and impossibility. The key was the notion of universal computation: the idea that a single universal Turing machine could be fed a finite program that would make it do anything that any Turing machine could do.

In a sense this meant that however sophisticated one's Turing machine technology might be, one would never be able to go beyond what any Turing machine that happened to be universal can do. And so if one asked a question, for example, about what the behavior of a Turing machine could be after an infinite time (say, does the machine ever reach a particular "halt" state), there might be no possible systematically finite way to answer that question, at least with any Turing machine.

But what about something other than a Turing machine?

Over the course of time, various other models of computational processes were proposed. But the surprising point that gradually emerged was that all the ones that seemed at all practical were ultimately equivalent. The original mathematical axiom system used in Gödel's theorem was also equivalent to a Turing machine. And so were all other reasonable models of what might constitute not only a computational process, but also a way to set up mathematics.

There may be some quite different way to set up a formal system than the way it is done in mathematics. But at least within mathematics as we currently define it, we can explicitly prove that there are impossibilities. We can prove that there are things that are genuinely infinite, and cannot meaningfully be reduced to something finite.

We know, for example (Hilbert's 10th problem), that there are polynomial equations involving integers where there is no finite mathematical procedure that will always determine whether the equations have solutions. It is not—as with the ordinary quintic equation—that with time some more sophisticated mathematical technology will be developed that allows solutions to be found. It is instead that within mathematics as an axiomatic system, it is simply impossible for there to be a finite general procedure.

So in mathematics there is in a sense "genuine impossibility."

Somewhat ironically, however, mathematics as a field of human activity tends to have little sense of this. And indeed there is a general belief in mathematics—much more so than in physics—that with time essentially any problem of "mathematical interest" will be solved.

A large part of the reason for this belief is that known examples of undecidable—or effectively impossible—problems tend to be complicated and contrived, and seem to have little to do with problems that could be of mathematical interest. My own work in exploring generalizations of mathematics gives strong evidence that undecidability is actually much closer at hand—and that in fact its apparent irrelevance is merely a reflection of the narrow historical path that mathematics as a field has followed. In a sense, the story is always the same—and to understand it sheds light on some of what might be impossible in physics. The issue is computational universality. Just where is the threshold for computational universality?

For once it is possible to achieve computational universality within a particular type of system or problem, it follows that the system or problem is in a sense as sophisticated as any other—and it is impossible to simplify it in any general way. And what I have found over and over again is that universality—and traces of it—occur in vastly simpler systems and problems than one might ever have imagined.

Indeed, my guess is that a substantial fraction of the famous unsolved problems in mathematics today are not unsolved because of a lack of mathematical technology—but because they are associated with universality, and so are fundamentally impossible to solve.

But what of physics?

Is there a direct correspondence of mathematical impossibility with physical impossibility? The answer is that it depends what physics is made of. If we can successfully reduce all of physics to mathematics, then mathematical impossibility in a sense becomes physical impossibility.

In the first few decades of the modern study of computation, the various models of computation that were considered were thought of mainly as representing processes—mechanical, electronic, or mathematical—that a human engineer or mathematician might set up. But particularly with the rise of models like cellular automata, the question increasingly arose of how these models—and computational processes they represent— might correspond to the actual operation of physics.

The traditional formulation of physics in terms of partial differential equations—or quantized fields—makes it difficult to see a correspondence. But the increasing implementation of physical models on computers has made the situation somewhat clearer.

There are two common technical issues. The first is that traditional physics models tend to be formulated in terms of continuous variables. The second is that traditional physics models tend not to say directly how a system should behave—but instead just to define an equation which gives a constraint on how the system should behave.

In modern times, good models of physical systems have often been found that are more obviously set up like traditional digital computations—with discrete variables, and explicit progression with time. But even traditional physical models are in many senses computational. For we know that even though there are continuous variables and equations to solve, there is an immense amount that we can work out about traditional physical models using, for example, Mathematica.

Mathematica obviously runs on an ordinary digital computer. But the point is that it can symbolically represent the entities in physical models. There can be a variable x that represents a continuous position, but to Mathematica it is just a finitely represented symbol, that can be manipulated using finite computational operations.

There are certainly questions that cannot obviously be answered by operating at a symbolic level—say about the precise location of some idealized particle represented by a real number. But when we imagine constructing an experiment or an apparatus, we specify it in a finite, symbolic way. And we might imagine that then we could answer all questions about its behavior by finite computational processes.

But this is undoubtedly not so. For it seems inevitable that within standard physical theories there is computational universality. And the result is that there will be questions that are impossible to answer in any finite way. Will a particular three-body gravitational system (or an idealized solar system) be stable forever? Or have some arbitrarily complicated form of instability?

Of course, it could be even worse.

If one takes a universal Turing machine, there are definite kinds of questions that cannot in general be answered about it—an example being whether it will ever reach a halt state from a given input. But at an abstract level, one can certainly imagine constructing a device that can answer such questions: doing some form of "hypercomputation." And it is quite straightforward to construct formal theories of whole hierarchies of such hypercomputations.

The way we normally define traditional axiomatic mathematics, such things are not part of it. But could they be part of physics? We do not know for sure. And indeed within traditional mathematical models of physics, it is a slippery issue.

In ordinary computational models like Turing machines, one works with a finite specification for the input that is given. And so it is fairly straightforward to recognize when some long and sophisticated piece of computational output can really be attributed to the operation of the system, and when it has somehow been slipped into the system through the initial conditions for the system.

But traditional mathematical models of physics tend to have parameters that are specified in terms of real numbers. And in the infinite sequence of digits in a precise real number, one can in principle pack all sorts of information—including, for example, tables of results that are beyond what a Turing machine can compute. And by doing this, it is fairly easy to set things up so that traditional mathematical models of physics appear to be doing hypercomputation.

But can this actually be achieved with anything like real, physical components?

I doubt it. For if one assumes that any device one builds, or any experiment one does, must be based on a finite description, then I suspect that it will never be possible to set up hypercomputation within traditional physical models.

In systems like Turing machines, there is a certain robustness and consistency to the notion of computation. Large classes of models, initial conditions, and other setups are equivalent at a computational level. But when hypercomputation is present, details of the setup tend to have large effects on the level of computation that can be reached, and there do not seem to be stable answers to questions about what is possible and not.

In traditional mathematical approaches to physics, we tend to think of mathematics as the general formalism, which in some special case applies to physics. But if there is hypercomputation in physics, it implies that in a sense we can construct physical tools that give us a new level of mathematics—and that answer problems in mathematics, though not by using the formalism of mathematics. And while at every level there are analogs of Gödel's theorem, the presence of hypercomputation in physics would in a sense overcome impossibilities in mathematics, for example giving us ways to solve all integer equations.

So could this be how our universe actually works?

From existing models in physics we do not know. And we will not ultimately know until we have a fundamental theory of physics.

Is it even possible to find a fundamental theory of physics? Again, we do not know for sure. It could be—a little like in hypercomputation—

that there will never be a finite description for how the universe works. But it is a fundamental observation—really the basis for all of natural science—that the universe does show order, and does appear to follow definite laws.

Is there in a sense some complete set of laws that provide a finite description for how the whole universe works? We will not know for sure until or unless we find that finite description—the ultimate fundamental theory.

One can argue about what that theory might be like. Is it perhaps finite, but very large, like the operating system of one of today's computers? Or is it not only finite, but actually quite small, like a few lines of computer code? We do not yet know.

Looking at the complexity and richness of the physical universe as we now experience it, we might assume that a fundamental theory—if it exists—would have to reflect all that complexity and richness, and itself somehow be correspondingly complex. But I have spent many years studying what is in effect a universe of possible theories—the computational universe of simple programs. And one of the clear conclusions is that in that computational universe it is easy to find immense complexity and richness, even among extremely short programs with extremely simple structures.

Will we actually be able to find our physical universe in this computational universe of possible universes? I am not sure. But certainly it is not obvious that we will not be able to do so. For already in my studies of the computational universe, I have found candidate universes that I cannot exclude as possible models of our physical universe.

If indeed there is a small ultimate model of our physical universe, it is inevitable that very few familiar features of our universe as we normally experience it will be visible in that model. For in a small model, there is in a sense no room to specify, say, the number of dimensions of space, the conservation of energy, or the spectrum of particles. Nor probably is there any room to have anything that corresponds directly to our normal notion of space or time.

Quite what the best representation for the model should be I am not sure. And indeed it is inevitable that there will be many seemingly quite different representations that only with some effort can be shown to be equivalent.

A particular representation that I have studied involves setting up a large number of nodes, connected in a network, and repeatedly updated according to some local rewrite rule. Within this representation, one can in effect just start enumerating possible universes, specifying their initial conditions and updating rules. Some candidate universes are very obviously not our physical universe. They have no notion of time, or no communication between different parts, or an infinite number of dimensions of space, or some other obviously fatal pathology.

But it turns out that there are large classes of candidate universes that already show remarkably suggestive features. For example, any universe that has a notion of time with a certain robustness property turns out in an appropriate limit to exhibit Special Relativity. And even more significantly, any universe that exhibits a certain conservation of finite dimensionality—as well as generating a certain level of effective microscopic randomness—will lead on a large scale to spacetime that follows Einstein's Equations for General Relativity.

It is worth emphasizing that the models I am discussing are in a sense much more complete than models one usually studies in physics. For traditionally in physics, it might be considered quite adequate to find equations, one of whose solutions successfully represents some feature of the universe. But in the models I have studied the concept is to have a formal system which starts from a particular initial state, then explicitly evolves so as to reproduce in every detail the precise evolution of our universe.

One might have thought that such a deterministic model would be excluded by what we know of quantum mechanics. But in fact the detailed nature of the model seems to make it quite consistent with quantum mechanics. And for example its network character makes it perfectly plausible to violate Bell's inequalities at the level of a large-scale limit of three-dimensional space.

So if in fact it turns out to be possible to find a model like this for our universe, what does it mean?

In some sense it reduces all of physics to mathematics. To work out what will happen in our universe becomes like working out the digits of pi: it just involves progressively applying some particular known algorithm.

Needless to say, if this is how things work, we will have immediately established that hypercomputation does not happen in our universe. And instead, only those things that are possible for standard computational systems like Turing machines can be possible in our universe.

But this does not mean that it is easy to know what is possible in our universe. For this is where the phenomenon of computational irreducibility comes in.

When we look at the evolution of some system—say a Turing machine or a cellular automaton—the system goes through some sequence of steps to determine its outcome. But we can ask whether perhaps there is some way to reduce the computational effort needed to find that outcome— some way to computationally reduce the evolution of the system.

And in a sense much of traditional theoretical physics has been based on the assumption that such computational reduction is possible. We want to find ways to predict how a system will behave, without having to explicitly trace each step in the actual evolution of the system.

But for computational reduction to be possible, it must in a sense be the case that the entity working out how a system will behave is computationally more sophisticated than the system itself.

In the past, it might not have seemed controversial to imagine that humans, with all their intelligence and mathematical prowess, would be computationally more sophisticated than systems in physics. But from my work on the computational universe, there is increasing evidence for a general Principle of Computational Equivalence, which implies that even systems with very simple rules can have the same level of computational sophistication as systems constructed in arbitrarily complex ways.

And the result of this is that many systems will exhibit computational irreducibility, so that their processes of evolution cannot be "outrun" by other systems—and in effect the only way to work out how the systems behave is to watch their explicit evolution.

This has many implications—not the least of which is that it can make it very difficult even to identify a fundamental theory of physics.

For let us say that one has a candidate theory—a candidate program for the universe. How can we find out whether that program actually is the program for our universe? If we just start running the program, we may quickly see that its behavior is simple enough that we can in effect computationally reduce it—and readily prove that it is not our universe.

But if the behavior is complex—and computationally irreducible—we will not be able to do this. And indeed as a practical matter in actually searching for a candidate model for our universe, this is a major problem. And all one can do is to hope that there is enough computational reducibility that one manages to identify known physical laws within the model universe.

It helps that if the candidate models for the universe are simple enough, then there will in a sense always be quite a distance from one model to another—so that successive models will tend to show very obviously different behavior. And this means that if a particular model reproduces any reasonable number of features of our actual universe, then there is a good chance that within the class of simple models, it will be essentially the only one that does so.

But, OK. Let us imagine that we have found an ultimate model for the universe, and we are confident that it is correct. Can we then work out what will be possible in the universe, and what will not?

Typically, there will be certain features of the universe that will be associated with computational reducibility, and for which we will readily be able to identify simple laws that define what is possible, and what is not.

Perhaps some of these laws will correspond to standard symmetries and invariances that have already been found in physics. But beyond these reducible features, there lies an infinite frontier of computational irreducibility. If we in effect reduce physics to mathematics, we still have to contend with phenomena like Gödel's theorem. So even given the underlying theory, we cannot work out all of its consequences.

If we ask a finite question, then at least in principle there will be a finite computational process to answer that question—though in practice we might be quite unable to run it. But to know what is possible, we also have to address questions that are in some sense not finite. Imagine that we want to know whether macroscopic spacetime wormholes are possible. It could be that we can use some computationally reducible feature of the universe to answer this.

But it could also be that we will immediately be confronted with computational irreducibility—and that our only recourse will for example be to start enumerating configurations of material in the universe to see if any of them end up evolving to wormholes. And it could even be that the question of whether any such configuration—of any size—exists could be formally undecidable, at least in an infinite universe.

But what about all those technologies that have been discussed in science fiction?

Just as we can imagine enumerating possible universes, so also we can imagine enumerating possible things that can be constructed in a particular universe. And indeed from our experience in exploring the computational universe of simple programs, we can expect that even simple constructions can readily lead to things with immensely rich and complex behavior.

But when do those things represent useful pieces of technology?

In a sense, the general problem of technology is to find things that can be constructed in nature, and then to match them with human purposes that they can achieve. And usually when we ask whether a particular type of technology is possible, what we are effectively asking is whether a particular type of human purpose can be achieved in practice. And to know this can be a surprisingly subtle matter, which depends almost as much on understanding our human context as it does on understanding features of physics.

Take for example almost any kind of transportation.

Earlier in human history, pretty much the only way to imagine that one would successfully achieve the purpose of transporting anything would be explicitly to move the thing from one place to another. But now there are many situations where what matters to us as humans is not the explicit material content of a thing, but rather the abstract information that represents it. And it is usually much easier to transport that information, often at the speed of light.

So when we say "will it ever be possible to get from here to there at a certain speed," we need to have a context for what would need to be transported. In the current state of human evolution, there is much that we do that can be represented as pure information, and readily transported. But we ourselves still have a physical presence, whose transportation seems like a different issue.

No doubt, though, we will one day master the construction of atomic-scale replicas from pure information. But more significantly, perhaps our very human existence will increasingly become purely informational—at which point the notion of transportation changes, so that just transporting information can potentially entirely achieve our human purposes.

There are different reasons for saying that things are impossible.

One reason is that the basic description of what should be achieved makes no sense. For example, if we ask "can we construct a universe where $2+2=5$?", then from the very meaning of the symbols in $2+2=5$, we can deduce that it can never be satisfied, whatever universe we are in.

There are other kinds of questions where at least at first the description seems to make no sense.

Like "is it possible to create another universe?" Well, if the universe is defined to be everything, then by definition the answer is obviously "no." But it is certainly possible to create simulations of other universes; indeed, in the computational universe of possible programs we can readily enumerate an infinite number of possible universes.

For us as physical beings, however, these simulations are clearly different from our actual physical universe. But consider a time in the future when the essence of the human condition has been transferred to purely informational form. At that time, we can imagine transferring our experience to some simulated universe, and in a sense existing purely within it—just as we now exist within our physical universe.

And from this future point of view, it will then seem perfectly possible to create other universes.

So what about time travel? There are also immediate definitional issues here. For at least if the universe has a definite history—with a single thread of time—the effect of any time travel into the past must just be reflected in the whole actual history that the universe exhibits.

We can often describe traditional physical models—for example for the structure of spacetime—by saying that they determine the future of a system from its past. But ultimately such models are just equations that connect different parameters of a system. And there may well be configurations of the system in which the equations cannot readily be seen just as determining the future from the past.

Quite which pathologies can occur with particular kinds of setups may well be undecidable, but when it seems that the future affects the past what is really being said is just that the underlying equations imply certain consistency conditions across time. And when one thinks of simple physical systems, such consistency conditions do not seem especially remarkable. But when one combines them with human experience—with its features of memory and progress—they seem more bizarre and paradoxical.

In some ancient time, one might have imagined that time travel for a person would consist of projecting them—or some aspect of them—far into the future. And indeed today when one sees writings and models that were constructed thousands of years ago for the afterlife, there is a sense in which that conception of time travel has been achieved.

And similarly, when one thinks of the past, the increasing precision with which molecular archaeology and the like can reconstruct things gives us something which at least at some time in history would have seemed tantamount to time travel.

Indeed, at an informational level—but for the important issue of computational irreducibility—we could reasonably expect to reconstruct the past and predict the future. And so if our human existence was purely informational, we would in some sense freely be able to travel in time.

The caveat of computational irreducibility is a crucial one, however, that affects the possibility of many kinds of processes and technologies.

We can ask, for example, whether it will ever be possible to do something like unscramble an egg, or in general in some sense to reverse time. The second law of thermodynamics has always suggested the impossibility of such things.

In the past, it was not entirely clear just what the fundamental basis for the second law might be. But knowing about computational irreducibility, we can finally see a solid basis for it. The basic idea is just that in many systems the process of evolution through time in effect so "encrypts" the information associated with the initial conditions for the system that no feasible measurement or other process can recognize what they were. So in effect, it would take a Maxwell's demon of immense computational power to unscramble the evolution.

In practice, however, as the systems we use for technology get smaller, and our practical powers of computation get larger, it is increasingly possible to do such unscrambling. And indeed that is the basis for a variety of important control systems and signal processing technologies that have emerged in recent years.

The question of just what kinds of effective reversals of time can be achieved by what level of technology depends somewhat on theoretical questions about computation. For example, if it is true that P!=NP, then certain questions about possible reversals will necessarily require immense computational resources.

There are many questions about what is possible that revolve around prediction.

Traditional models in physics tend to deny the possibility of prediction for two basic reasons. The first is that the models are usually assumed to be somehow incomplete, so that the systems they describe are subject to unknown—and unpredictable—effects from the outside. The second reason is quantum mechanics—which in its traditional formulation is fundamentally probabilistic.

Quite what happens even in a traditional quantum formulation when one tries to describe a whole sequence from the construction of an experiment to the measurement of its results has never been completely clear. And for example it is still not clear whether it is possible to generate a perfectly random sequence—or whether in effect the operation of the preparation and measurement apparatus will always prevent this. But even if—as in candidate models of fundamental physics that I have investigated—there is no ultimate randomness in quantum mechanics, there is still another crucial barrier to prediction: computational irreducibility.

One might have thought that in time there would be some kind of acceleration in intelligence that would allow our successors to predict anything they want about the physical universe.

But computational irreducibility implies that there will always be limitations. There will be an infinite number of pockets of reducibility where progress can be made. But ultimately the actual evolution of the universe in a sense achieves something irreducible—which can only be observed, not predicted.

What if perhaps there could be some collection of extraterrestrial intelligences around the universe who combine to try to compute the future of the universe?

We are proud of the computational achievements of our intelligence and our civilization. But what the Principle of Computational Equivalence implies is that many processes in nature are ultimately equivalent in their computational sophistication. So in a sense the universe is already as intelligent as we are, and whatever we develop in our technology cannot overcome that. It is only that with our technology we guide the universe in ways that we can think of as achieving our particular purposes.

However, if it turns out—as I suspect—that the whole history of the universe is determined by a particular, perhaps simple, underlying rule, then we are in a sense in an even more extreme situation.

For there is in a sense just one possible history for the universe. So at some level this defines all that is possible. But the point is that to answer specific questions about parts of this history requires irreducible computational work—so that in a sense perhaps there can still be essentially infinite amounts of surprise about what is possible, and we can still perceive that we act with free will.

So what will the limit of technology in the future be like?

Today almost all the technology we have has been created through traditional methods of engineering: by building up what is needed one step at a time, always keeping everything simple enough that we can foresee what the results will be.

But what if we just searched the computational universe for our technology? One of the discoveries from exploring the computational universe is that even very simple programs can exhibit rich and complex behavior. But can we use this for technology?

The answer, it seems, is often yes. The methodology for doing this is not yet well known. But in recent years my own technology development projects have certainly made increasingly central use of this approach.

One defines some particular objective—say generating a hash code, evaluating a mathematical function, creating a musical piece, or recognizing a class of linguistic forms. Then one searches the computational universe for a program that achieves the objective. It might be that the simplest program that would be needed would be highly complex—and out of reach of enumerative search methods. But the Principle of Computational Equivalence suggests that this will tend not to be the case—and in practice it seems that it is not.

And indeed one often finds surprisingly simple programs that achieve all sorts of complex purposes.

Unlike things created by traditional engineering, however, there is no constraint that these programs operate in ways that we as humans can readily understand. And indeed it is common to find that they do not. Instead, in a sense, they tend to operate much more like many systems in nature—that we can describe as achieving a certain overall purpose, but can't readily understand how they do it.

Today's technology tends at some level to look very regular— to exhibit simple geometrical or informational motifs, like rotary motion or iterative execution. But technology that is "mined" from the computational universe will usually not show such simplicity. It will look much more like many systems in nature—and operate in a sense much more efficiently with its resources, and much closer to computational irreducibility.

The fact that a system can be described as achieving some particular purpose by definition implies a certain computational reducibility in its behavior.

But the point is that as technology advances, we can expect to see less and less computational reducibility that was merely the result of engineering or historical development—and instead to see more and more perfect computational irreducibility.

It is in a sense a peculiar situation, forced on us by the Principle of Computational Equivalence. We might have believed that our own intelligence, our technology, and the physical universe we inhabit would all have different levels of computational sophistication.

But the Principle of Computational Equivalence implies that they do not. So even though we may strive mightily to create elaborate technology, we will ultimately never be able give it any fundamentally greater level of computational sophistication. Indeed, in a sense all we will ever be able to do is to equal what already happens in nature.

And this kind of equivalence has fundamental implications for what we will consider possible.

Today we are in the early stages of merging our human intelligence and existence with computation and technology. But in time this merger will no doubt be complete, and our human existence will in a sense be played out through our technology. Presumably there will be a progressive process of optimization—so that in time the core of our thoughts and activities will simply consist of some complicated patterns of microscopic physical effects.

But looking from outside, a great many systems in nature similarly show complicated patterns of microscopic physical effects. And what the Principle of Computational Equivalence tells us is that there can ultimately be no different level of computational sophistication in the effects that are the result of all our civilization and technology development—and effects that just occur in nature.

We might think that processes corresponding to future human activities would somehow show a sense of purpose that would not be shared by processes that just occur in nature. But in the end, what we define as purpose is ultimately just a feature of history—defined by the particular details of the evolution of our civilization.

We can certainly imagine in some computational way enumerating all possible purposes—just as we can imagine enumerating possible computational or physical or biological systems. So far in human history we have pursued only a tiny fraction of all possible purposes. And perhaps the meaningful future of our civilization will consist only of pursuing some modest extrapolation of what we have pursued so far.

So which of our purposes can we expect to achieve in the physical universe? The answer, I suspect, is that once our existence is in effect purely computational, we will in a sense be able to program things so as to achieve a vast range of purposes. Today we have a definite, fixed physical existence. And to achieve a purpose in our universe we must mold physical components to achieve that purpose. But if our very existence is in effect purely computational, we can expect not only to mold the outside physical universe, but also in a sense to mold our own computational construction.

The result is that what will determine whether a particular purpose can be achieved in our universe will more be general abstract issues like computational irreducibility than issues about the particular physical laws of our universe. And there will certainly be some purposes that we can in principle define, but which can never be achieved because they require infinite amounts of irreducible computation.

In our science, technology, and general approach to rational thinking, we have so far in our history tended to focus on purposes which are not made impossible by computational irreducibility—though we may not be able to see how to achieve them with physical components in the context of our current existence. As we extrapolate into the future of our civilization, it is not clear how our purposes will evolve—and to what extent they will become enmeshed with computational irreducibility, and therefore seem possible or not.

So in a sense what we will ultimately perceive as possible in physics depends more on the evolution of human purposes than it does on the details of the physical universe. In some ways this is a satisfying result.

For it suggests that we will ultimately never be constrained in what we can achieve by the details of our physical universe. The constraints on our future will not be ones of physics, but rather ones of a deeper nature. It will not be that we will be forced to progress in a particular direction because of the specific details of the particular physical universe in which we live. But rather—in what we can view as an ultimate consequence of the Principle of Computational Equivalence—the constraints on what is possible will be abstract features of the general properties of the computational universe. They will not be a matter of physics—but instead of the general science of the computational universe.

My Life in Technology—As Told at the Computer History Museum

April 19, 2016

I normally spend my time trying to build the future. But I find history really interesting and informative, and I study it quite a lot. Usually it's other people's history. But the Computer History Museum asked me to talk today about my own history, and the history of technology I've built. So that's what I'm going to do here.

This happens to be a really exciting time for me—because a bunch of things that I've been working on for more than 30 years are finally coming to fruition. And mostly that's what I've been out in the Bay Area this week talking about.

The focus is the Wolfram Language, which is really a new kind of language—a knowledge-based language—in which as much knowledge as possible about computation and about the world is built in, and in which the language automates as much as possible so one can go as directly as possible from computational thinking to actual implementation.

And what I want to do here is to talk about how all this came to be, and how things like Mathematica and Wolfram|Alpha emerged along the way.

Inevitably a lot of what I'm going to talk about is really my story: basically the story of how I've spent most of my life so far building a big stack of technology and science. When I look back, some of what's happened seems sort of inevitable and inexorable. And some I didn't see coming.

But let me begin at the beginning. I was born in London, England, in 1959—so, yes, I'm outrageously old, at least by my current standards. My father ran a small company—doing international trading of textiles—for nearly 60 years, and also wrote a few "serious fiction" novels. My mother was a philosophy professor at Oxford. I actually happened to notice her textbook on philosophical logic in the Stanford bookstore last time I was there.

You know, I remember when I was maybe five or six being bored at some party with a bunch of adults, and somehow ending up talking at great length to some probably very distinguished Oxford philosopher—who I heard say at the end, "One day that child will be a philosopher—but it may take a while." Well, they were right. It's sort of funny how these things work out.

Here's me back then:

I went to elementary school in Oxford—to a place called the Dragon School, that I guess happens to be probably the most famous elementary school in England. Wikipedia seems to think the most famous people now from my class there are myself and the actor Hugh Laurie.

Here's one of my school reports, from when I was seven. Those are class ranks. So, yes, I did well in poetry and geography, but not in math. (And, yes, it's England, so they taught "Bible Study" in school, at least then.) But at least it said "He is full of spirit & determination; he should go far"....

But OK, that was 1967, and I was learning Latin and things—but what I really liked was the future. And the big future-oriented thing happening back then was the space program. And I was really interested in that, and started collecting all the information I could about every spacecraft launched—and putting together little books summarizing it. And I discovered that even from England one could write to NASA and get all this great stuff mailed to one for free.

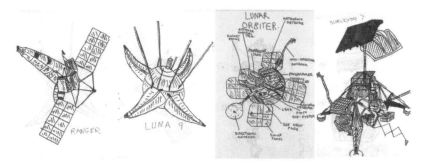

Well, back then, there was supposed to be a Mars colony any day, and I started doing little designs for that, and for spacecraft and things.

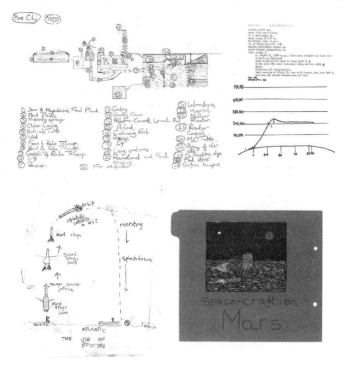

And that got me interested in propulsion and ion drives and stuff like that—and by the time I was 11 what I was really interested in was physics.

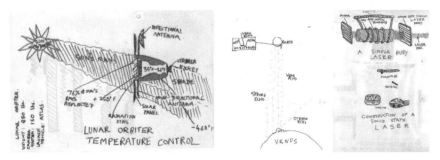

And I discovered—having nothing to do with school—that if one just reads books one can learn stuff pretty quickly. I would pick areas of physics and try to organize knowledge about them. And when I was turning 12 I ended up spending the summer putting together all the facts I could accumulate about physics. And, yes, I suppose you could call some of these "visualizations." And, yes, like so much else, it's on the web now:

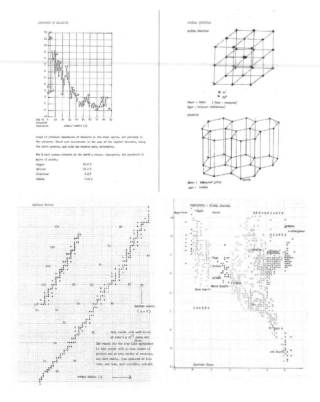

I found this again a few years ago—around the time Wolfram|Alpha came out—and I thought, "Oh my gosh, I've been doing the same thing all my life!" And then of course I started typing in numbers from when I was 11 or 12 to see if Wolfram|Alpha got them right. It did, of course:

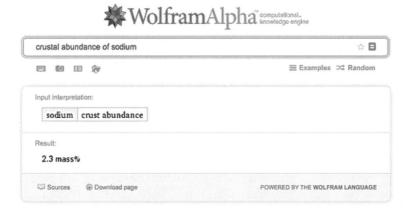

Well, when I was 12, following British tradition I went to a so-called public school that's actually a private school. I went to the most famous such school—Eton—which was founded about 50 years before Columbus came to America. And, oh so impressively ☺, I even got the top scholarship among new kids in 1972.

Yes, everyone wore tailcoats all the time, and King's Scholars, like me, wore gowns too—which provided excellent rain protection etc. I think I avoided these annual Harry Potter–like pictures all but one time:

And back in those Latin-and-Greek-and-tailcoat days I had a sort of double life, because my real passion was doing physics.

The summer when I turned 13 I put together a summary of particle physics:

And I made the important meta-discovery that even if one was a kid, one could discover stuff. And I started just trying to answer questions about physics, either by finding answers in books, or by figuring them out myself. And by the time I was 15 I started publishing papers about physics. Yes, nobody asks how old you are when you mail a paper in to a physics journal.

But, OK, something important for me had happened back when I was 12 and first at Eton: I got to know my first computer. It's an Elliott 903C. This is not the actual one I used, but it's similar:

It had come to Eton through a teacher of mine named Norman Routledge, who had been a friend of Alan Turing's. It had 8 kilowords of 18-bit ferrite core memory, and you usually programmed it with paper— or Mylar—tape, most often in a little 16-instruction assembler called SIR.

It often seemed like one of the most important skills was rewinding the tape as quickly as possible after it got dumped in a bin after going through the optical reader.

Anyway, I wanted to use the computer to do physics. When I was 12 I had gotten this book:

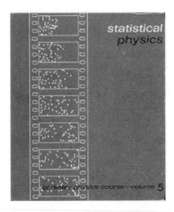

What's on the cover is supposed to be a simulation of gas molecules showing increasing randomness and entropy. As it happens, years later I discovered this picture was actually kind of a fake. But back when I was 12, I really wanted to reproduce it—with the computer.

It wasn't so easy. The molecule positions were supposed to be real numbers; one had to have an algorithm for collisions; and so on. And to make this fit on the Elliott 903 I ended up simplifying a lot—to what was actually a 2D cellular automaton.

Well, a decade after that, I made some big discoveries about cellular automata. But back then I was unlucky with my cellular automaton rule, and I ended up not discovering anything with it. And in the end my biggest achievement with the Elliott 903 was writing a punched tape loader for it.

You see, the big problem with the Mylar tape that one used for serious programs is that it would get statically electrically charged and pick up little confetti holes, so the bits would be read wrong. Well, for my loader, I came up with what I later found out were error-correcting codes—and I set it up so that if the checks failed, the tape would stop in the reader, and you could pull it back a couple of feet, and then reread it, after shaking out the confetti.

OK, so by the time I was 16 I had published some physics papers and was starting to be known in physics circles—and I left school, and went to work at a British government lab called the Rutherford Lab that did particle physics research.

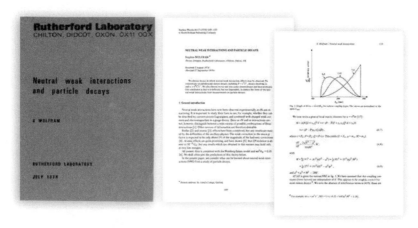

Now you might remember from my age-seven school report that I didn't do very well in math. Things got a bit better when I started to use a slide rule, and then in 1972 a calculator—of which I was a very early adopter. But I never liked doing school math, or math calculations in general. Well, in particle physics there's a lot of math to be done—and so my dislike of it was a problem.

At the Rutherford Lab, two things helped. First, a lovely HP desktop computer with a plotter, on which I could do very nice interactive computations. And second, a mainframe for crunchier things, that I programmed in Fortran.

Well, after my time at the Rutherford Lab I went to college at Oxford. Within a very short time I'd decided this was a mistake—but in those days one didn't actually have to go to lectures for classes—so I was able to just hide out and do physics research. And mostly I spent my time in a nice underground air-conditioned room in the Nuclear Physics building—that had terminals connected to a mainframe, and to the ARPANET.

And that was when—in 1976—I first started using computers to do symbolic math, and algebra and things. Feynman diagrams in particle physics involve lots and lots of algebra. And back in 1962, I think, three physicists had met at CERN and decided to try to use computers to do this. They had three different approaches. One wrote a system called ASHMEDAI in Fortran. One—influenced by John McCarthy at Stanford—wrote a system called Reduce in Lisp. And one wrote a system called SCHOONSCHIP in CDC 6000 series assembly language, with mnemonics in Dutch. Curiously, years later, one of these physicists won a Nobel Prize. It was Tini Veltman—the one who wrote SCHOONSCHIP in assembly language.

Anyway, back in 1976 very few people other than the creators of these systems used them. But I started using all of them. But my favorite was a quite different system, written in Lisp at MIT since the mid-1960s. It was a system called Macsyma. It ran on the Project MAC PDP-10 computer. And what was really important to me as a 17-year-old kid in England was that I could get to it on the ARPANET.

It was host 236. So I would type @O 236, and there I was in an interactive operating system. Someone had taken the login SW. So I became Swolf, and started to use Macsyma.

I spent the summer of 1977 at Argonne National Lab—where they actually trusted physicists to be right in the room with the mainframe.

Then in 1978 I went to Caltech as a graduate student. By that point, I think I was the world's largest user of computer algebra. And it was so neat, because I could just compute all this stuff so easily. I used to have fun putting incredibly ornate formulas in my physics papers. Then I could see if anyone was reading the papers, because I'd get letters saying, "How did you derive line such-and-such from the one before?"

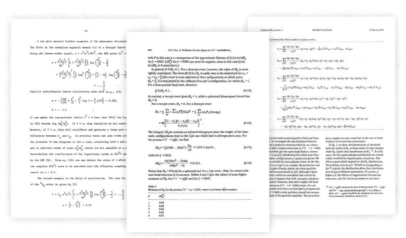

I got a reputation for being a great calculator. Which was of course 100% undeserved—because it wasn't me, it was just the computer. Well, actually, to be fair, there was part that was me. You see, by being able to compute so many different examples, I had gotten a new kind of intuition. I was no good at computing integrals myself, but I could go back and forth with the computer, knowing from intuition what to try, and then doing experiments to see what worked.

I was writing lots of code for Macsyma, and building this whole tower. And sometime in 1979 I hit the edge. Something new was needed. (Notice, for example, the ominous "MACSYMA RELOAD" line in the left-hand image.)

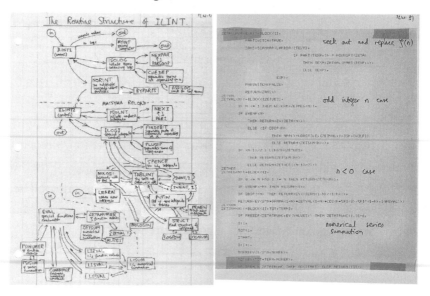

Well, in November 1979, just after I turned 20, I put together some papers, called it a thesis, and got my PhD. And a couple of days later I was visiting CERN in Geneva—and thinking about my future in, I thought, physics. And the one thing I was sure about was that I needed something beyond Macsyma that would let me compute things. And that was when I decided I had to build a system for myself. And right then and there, I started designing the system, handwriting its specification.

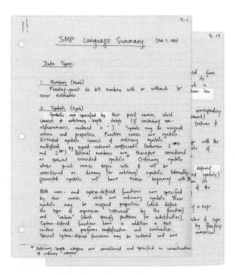

At first it was going to be ALGY—The Algebraic Manipulator. But I quickly realized that I actually had to make it do much more than algebraic manipulation. I knew most of the general-purpose computer languages of the time—both the ALGOL-like ones, and ones like Lisp and APL. But somehow they didn't seem to capture what I wanted the system to do.

So I guess I did what I'd learned in physics: I tried to drill down to find the atoms—the primitives—of what was going on. I knew a certain amount about mathematical logic, and the history of attempts to formulate things using logic and so on—even if my mother's textbook about philosophical logic didn't exist yet.

The whole history of this effort at formalization—through Aristotle, Leibniz, Frege, Peano, Hilbert, Whitehead, Russell, and so on—is really interesting. But that's a different talk. Back in 1979 it was thinking about this kind of thing that led me to the design I came up with, that was based on the idea of symbolic expressions, and doing transformations on them.

I named what I wanted to build SMP: a Symbolic Manipulation Program, and started recruiting people from around Caltech to help me with it. Richard Feynman came to a bunch of the meetings I had to discuss the design of SMP, offering various ideas—which I have to admit I considered hacky—about shortcuts for interacting with the system. Meanwhile, the physics department had just gotten a VAX 11/780, and after some wrangling, it was made to run Unix. Meanwhile, a young physics grad student named Rob Pike—more recently creator of the Go programming language—persuaded me that I should write the code for my system in the "language of the future": C.

I got pretty good at writing C, for a while averaging about a thousand lines a day. And with the help of a somewhat colorful collection of characters, by June 1981, the first version of SMP existed—with a big book of documentation I'd written.

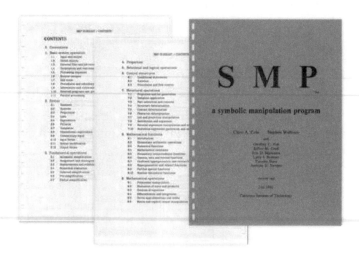

OK, you might ask: so can we see SMP? Well, back when we were working on SMP I had the bright idea that we should protect the source code by encrypting it. And—you guessed it—over a span of three decades nobody remembers the password. And until a little while ago, that was the situation.

In another bright idea, I had used a modified version of the Unix crypt program to do the encryption—thinking that would be more secure. Well, as part of the 25th anniversary of Mathematica a couple of years ago, we did a crowdsourced project to break the encryption—and we did it. Unfortunately it wasn't easy to compile the code though—but thanks to a 15-year-old volunteer, we've actually now got something running.

So here it is: running inside a VAX virtual machine emulator, I can show you for the first time in public in 30 years—a running version of SMP.

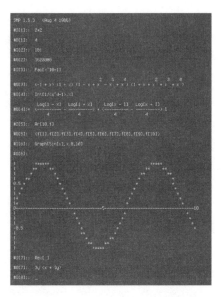

SMP had a mixture of good ideas, and very bad ideas. One example of a bad idea—actually suggested to me by Tini Veltman, author of SCHOONSHIP—was representing rationals using floating point, so one could make use of the faster floating-point instructions on many processors. But there were plenty of other bad ideas too, like having a garbage collector that had to crawl the stack and realign pointers when it ran.

There were some interesting ideas. Like what I called "projections"—which were essentially a unification of functions and lists. They were almost wonderful, but there were confusions about currying—or what I called tiering. And there were weird edge cases about things that were almost vectors with sequential integer indices.

But all in all, SMP worked pretty well, and I certainly found it very useful. So now the next problem was what to do with it. I realized it needed a real team to work on it, and I thought the best way to get that was somehow to make it commercial. But at the time I was a 21-year-old physics-professor type, who didn't know anything about business.

So I thought, let me go to the tech transfer office at the university, and ask them what to do. But it turned out they didn't know, because, as they explained, "Mostly professors don't come to us; they just start their own companies." "Well," I said, "can I do that?" And right then and there the lawyer who pretty much was the tech transfer office pulled out the faculty handbook, and looked through it, and said, "Well, yes, it says copyrightable materials are owned by their authors, and software is copyrightable, so, yes, you can do whatever you want."

And so off I went to try to start a company. Though it turned out not to be so simple—because suddenly the university decided that actually I couldn't just do what I wanted.

A couple of years ago I was visiting Caltech and I ran into the 95-year-old chap who had been the provost at the time—and he finally filled in for me the remaining details of what he called the "Wolfram Affair." It was more bizarre than one could possibly imagine. I won't tell it all here. But suffice it to say that the story starts with Arnold Beckman, Caltech postdoc in 1929, claiming rights to the pH meter, and starting Beckman Instruments—and then in 1980 being chairman of the Caltech board of trustees and being upset when he realized that gene-sequencing technology had been invented at Caltech and had "walked off campus" to turn into Applied Biosystems.

But the company I started weathered this storm—even if I ended up quitting Caltech, and Caltech ended up with a weird software-ownership policy that affected their computer-science recruiting efforts for a long time.

I didn't do a great job starting what I called Computer Mathematics Corporation. I brought in a person—who happened to be twice my age—to be CEO. And rather quickly things started to diverge from what I thought made sense.

One of my favorite moments of insanity was the idea to get into the hardware business and build a workstation to run SMP on. Well, at the time no workstation had enough memory, and the 68000 didn't handle virtual memory. So a scheme was concocted whereby two 68000s would run an instruction out of step, and if the first one saw a page fault, it would stop the other one and fetch the data. I thought it was nuts. And I also happened to have visited Stanford, and run into a grad student named Andy Bechtolsheim who was showing off a Stanford University Network—SUN—workstation with a cardboard box as a case.

But worse than all that, this was 1981, and there was the idea that AI—in the form of expert systems—was hot. So the company merged with another company that did expert systems, to form what was called Inference Corporation (which eventually became Nasdaq:INFR). SMP was the cash cow—selling for about $40,000 a copy to industrial and government research labs. But the venture capitalists who'd come in were convinced that the future was expert systems, and after not very long, I left.

Meanwhile I'd become a big expert on the intellectual property policies of universities—and eventually went to work at the Institute for Advanced Study in Princeton, where the director very charmingly said that since they'd "given away the computer" after von Neumann died, it didn't make much sense for them to claim IP rights to anything now.

I dived into basic science, working a lot on cellular automata, and discovering some things I thought were very interesting. Here's me with my SUN workstation with cellular automata running on it (and, yes, the mollusc looks like the cellular automaton):

I did some consulting work, mostly on technology strategy, which was very educational, particularly in seeing things not to do. I did quite a lot of work for Thinking Machines Corporation. I think my most important contribution was going to see the movie *WarGames* with Danny Hillis—and as we were walking out of the movie theater, saying to Danny, "Maybe your computer should have flashing lights too." (The flashing lights ended up being a big feature of the Connection Machine computer—certainly important in its afterlife in museums.)

I was mostly working on basic science—but "because it would be easy" I decided to do a software project of building a C interpreter that we called IXIS. I hired some young people—one of whom was Tsutomu Shimomura, whom I'd already fished out of several hacking disasters. I made the horrible mistake of writing the boring code nobody else wanted to write myself—so I wrote a (quite lovely) text editor, but the whole project flopped.

I had all kinds of interactions with the computer industry back then. I remember Nathan Myhrvold, then a physics grad student at Princeton, coming to see me to ask what to do with a window system he'd developed. My basic suggestion was "sell it to Microsoft." As it happens, Nathan later became CTO of Microsoft.

Well, by about 1985 I'd done a bunch of basic science I was pretty pleased with, and I was trying to use it to start the field of what I called complex systems research. I ended up getting a little involved in an outfit called the Rio Grande Institute—that later became the Santa Fe Institute—and encouraging them to pursue this kind of research. But I wasn't convinced about their chances, and I resolved to start my own research institute.

So I went around to lots of different universities, in effect to get bids. The University of Illinois won, ironically in part because they thought it would help their chances getting funding from the Beckman Foundation—which in fact it did. So in August 1986, off I went to the University of Illinois, and the cornfields of Champaign-Urbana, 100 miles south of Chicago.

I think I did pretty well at recruiting faculty and setting things up for the new Center for Complex Systems Research—and the university lived up to its end of the bargain too. But within a few weeks I started to think it was all a big mistake. I was spending all my time managing things and trying to raise money—and not actually doing science.

So I quickly came up with Plan B. Rather than getting other people to help with the science I wanted to do, I would set things up so I could just do the science myself, as efficiently as possible. And this meant two things: first, I had to have the best possible tools; and second, I needed the best possible environment for myself.

When I was doing my basic science I kept on using different tools. There was some SMP. Quite a lot of C. Some PostScript, and graphics libraries, and things. And a lot of my time was spent gluing all this stuff together. And what I decided was that I should try to build a single system that would just do all the stuff I wanted to do—and that I could expect to keep growing forever.

Well, meanwhile, personal computers were just getting to the point where it was plausible to build a system like this that would run on them. And I knew a lot about what to do—and not do—from my experience with SMP. So I started designing and building Mathematica.

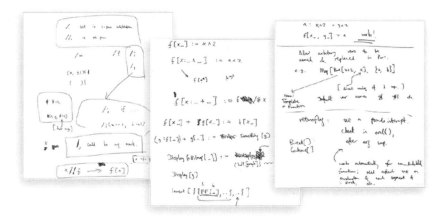

My scheme was to write documentation to define what to build. I wrote a bunch of core code—for example for the pattern matcher—a surprising amount of which is still in the system all these years later. The design of Mathematica was in many respects less radical and less extreme than SMP. SMP had insisted on using the idea of transforming symbolic expressions for everything—but in Mathematica I saw my goal as being to design a language that would effectively capture all the possible different paradigms for thinking about programming in a nice seamless way.

At first, of course, Mathematica wasn't called Mathematica. In a strange piece of later fate, it was actually called Omega. It went through other names. There was Polymath. And Technique. Here's a list of names. It's kind of shocking to me how many of these—even the really horrible ones—have actually been used for products in the years since.

```
Names.do        Mon Aug 31 16:50:14 1987

Some perhaps possible product names:

Technique
Hypermath
Polymath
M
Omega
QED
Axiom
Formula
Exact
Xi
X
Synthesis
Extrapolate
Insight
Metacalc
Mathtool
Paramath
Mathbox
Mathesis
Mathdriver
Mathscript
Mathica
Macala
Exacal
Axian

Some impossible product names:

Mathesourus
Igor
Mathdroid
```

Well, meanwhile, I was starting to investigate how to build a company around the system. My original model was something like what Adobe was doing at the time with PostScript: we build core IP, then license it to hardware companies to bundle. And as it happened, the first person to show interest in that was Steve Jobs, who was then in the middle of doing NeXT.

Well, one of the consequences of interacting with Steve was that we talked about the name of the product. With all that Latin I'd learned in school, I'd thought about the name "Mathematica" but I thought it was too long and ponderous. Steve insisted that "that's the name"—and had a whole theory about taking generic words and romanticizing them. And eventually he convinced me.

It took about 18 months to build Version 1 of Mathematica. I was still officially a professor of physics, math, and computer science at the University of Illinois. But apart from that I was spending every waking hour building software and later making deals.

We closed a deal with Steve Jobs at NeXT to bundle Mathematica on the NeXT computer:

We also made a bunch of other deals. With Sun, through Andy Bechtolsheim and Bill Joy. With Silicon Graphics, through Forest Baskett. With Ardent, through Gordon Bell and Cleve Moler. With the AIX/RT part of IBM, basically through Andy Heller and Vicky Markstein.

And eventually we set a release date: June 23, 1988.

Meanwhile, as documentation for the system, I wrote a book called *Mathematica: A System for Doing Mathematics by Computer*. It was going to be published by Addison-Wesley, and it was the longest lead-time element of the release. And it ended up being very tight, because the book was full of fancy PostScript graphics—which nobody could apparently figure out how to render at high-enough resolution. So eventually I just took a hard disk to a friend of mine in Canada who had a phototypesetting company, and he and I babysat his phototypesetting machine over a holiday weekend, after which I flew to Logan Airport in Boston and handed the finished film for the book to a production person from Addison-Wesley.

We decided to do the announcement of Mathematica in Silicon Valley, and specifically at the TechMart place in Santa Clara. In those days Mathematica couldn't run under MS-DOS because of the 640K memory limit. So the only consumer version was for the Mac. And the day before the announcement there we were stuffing disks into boxes, and delivering them to the ComputerWare software store in Palo Alto.

The announcement was a nice affair. Steve Jobs came—even though he was not really "out in public" at the time. Larry Tesler came from Apple—courageously doing a demo himself. John Gage from Sun had the sense to get all the speakers to sign a book:

And so that was how Mathematica was launched. *The Mathematica Book* became a bestseller in bookstores, and from that people started understanding how to use Mathematica. It was really neat seeing all these science types and so on—of all ages—who'd basically never used computers themselves before, starting to just compute things themselves.

It was fun looking through registration cards. Lots of interesting and famous names. Sometimes some nice juxtapositions. Like when I'd just seen an article about Roger Penrose and his new book in *Time* magazine with the headline "Those Computers Are Dummies"... but then there was Roger's registration card for Mathematica.

As part of the growth of Mathematica, we ended up interacting with pretty much all possible computer companies, and collected all kinds of exotic machines. Sometimes that came in handy, like when the Morris worm came through the internet, and our gateway machine was a weird Sony workstation with a Japanese OS that the worm hadn't been built for.

There were all kind of porting adventures. Probably my favorite was on the Cray-2. With great effort we'd gotten Mathematica compiled. And there we were, ready for the first calculation. And someone typed 2+2. And—I kid you not—it came out "5". I think it was an issue with integer vs. floating-point representation.

You know, here's a price list from 1990 that's a bit of a stroll down computer memory lane:

We got a boost when the NeXT computer came out, with Mathematica bundled on it. I think Steve Jobs made a good deal there, because all kinds of people got NeXT machines to run Mathematica. Like the Theory group at CERN—where the systems administrator was Tim Berners-Lee, who decided to do a little networking experiment on those machines.

Well, a couple of years in, the company was growing nicely—we had maybe 150 employees. And I thought to myself: I built this because I wanted to have a way to do my science, so isn't it time I started doing that?

Also, to be fair, I was injecting new ideas at too high a rate; I was worried the company might just fly apart. But anyway, I decided I would take a partial sabbatical—for maybe six months or a year—to do basic science and write a book about it.

So I moved from Illinois to the Oakland Hills—right before the big fire there, which narrowly missed our house. And I started being a remote CEO—using Mathematica to do science. Well, the good news was that I started discovering lots and lots of science. It was kind of a "turn a telescope to the sky for the first time" moment—except now it was the computational universe of possible programs.

It was really great. But I just couldn't stop—because there kept on being more and more things to discover. And all in all I kept on doing it for ten and a half years. I was really a hermit, mostly living in Chicago, and mostly interacting only virtually... although my oldest three children were born during that period, so there were humans around!

I had thought maybe there'd be a coup at the company. But there wasn't. And the company continued to steadily grow. We kept on doing new things.

Here's our first website, from October 7, 1994:

And it wasn't too long after that we started doing computation on the web:

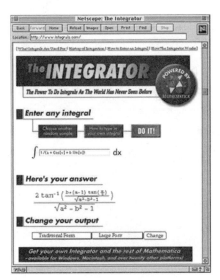

I actually took a break from my science in 1996 to finish a big new version of Mathematica. Back in 1988 lots of people used Mathematica through a command-line interface. In fact, it's still there today. 1989^{1989} is the basic computation I've been using since, yes, 1989, to test speed on a new machine. And actually a basic Raspberry Pi today gives a pretty good sense of what it was like back at the beginning.

But, OK, on the Mac and on NeXT back in 1988 we'd invented these things we called notebooks that were documents that mixed text and graphics and structure and computation—and that was the UI. It was all very modern, with a clean front-end/kernel architecture where it was easy to run the kernel on a remote machine—and by 1996 a complete symbolic XML-like representation of the structure of the notebooks.

Maybe I should say something about the software engineering of Mathematica. The core code was written in an extension of C— actually an object-oriented version of C that we had to develop ourselves, because C++ wasn't efficient enough back in 1988. Even from the beginning, some code was written in the Mathematica top-level language—that's now the Wolfram Language—and over the years a larger and larger fraction of the code was that way.

Well, back at the beginning it was very challenging getting the front end to run on different machines. And we wound up with different codebases on Mac, NeXT, Microsoft Windows, and X Windows. And in 1996 one of the achievements was merging all that together. And for almost 20 years the code was gloriously merged—but now we've again got separate codebases for desktop, browser, and mobile, and history is repeating itself.

Back in 1996 we had all kinds of ways to get the word out about the new Mathematica Version 3. My original Mathematica book had now become quite large, to accommodate all the things we were adding.

And we had a couple of other "promotional vehicles" that we called the MathMobiles that drove around with the latest gear inside—and served as moving billboard ads for our graphics.

There were Mathematicas everywhere, getting used for all kinds of things. And of course wild things sometimes happened. Like in 1997 when Mike Foale had a PC running Mathematica on the Mir space station. Well, there was an accident, and the PC got stuck in a part of the space station that got depressurized. Meanwhile, the space station was tumbling, and Mike was trying to debug it—and wanted to use Mathematica to do it. So he got a new copy on the next supply mission—and installed it on a Russian PC.

But there was a problem. Because our DRM system immediately said, "That's a Russian PC; you can't run a US-licensed Mathematica there!" And that led to what might be our all-time most exotic customer service call: "The user is in a tumbling space station." But fortunately we could just issue a different password—Mike solved the equations, and the space station was stabilized.

Well, after more than a decade—in 2002—I finally finished my science project and my big book:

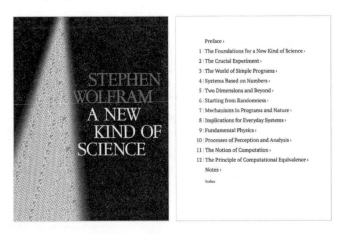

During my "science decade" the company had been steadily growing, and we'd built up a terrific team. But not least because of things I'd learned from my science, I thought it could do more. It was refreshing coming back to focus on it again. And I rather quickly realized that the structure we'd built could be applied to lots of new things.

Math had been the first big application of Mathematica, but the symbolic language I'd built was much more general than that. And it was pretty exciting seeing what we could do with it. One of the things in 2006 was representing user interfaces symbolically, and being able to create them computationally. And that led for example to CDF (our Computable Document Format), and things like our Wolfram Demonstrations Project.

We started doing all sorts of experiments. Many went really well. Some went a bit off track. We wanted to make a poster with all the facts we knew about mathematical functions. First it was going to be a small poster, but then it became 36 feet of poster... and eventually the Wolfram Functions Site, with 300,000+ formulas:

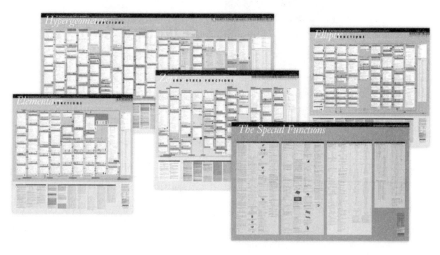

It was the time of the cellphone ringtone craze, and I wanted a personal ringtone. So we came up with a way to use cellular automata to compose an infinite variety of ringtones, and we put it on the web. It was actually an interesting AI-creativity experience, and music people liked it. But after messing around with phone carriers for six months, we pretty much didn't sell a single ringtone.

But, anyway, having for many years been a one-product company making Mathematica, we were starting to get the idea that we could not only add new things to Mathematica—but also invent all kinds of other stuff.

Well, I mentioned that back when I was a kid I was really interested in trying to do what I'd now call "making knowledge computable": take the knowledge of our civilization and build something that could automatically compute answers to questions from it. For a long time I'd assumed that to do that would require making some kind of brain-like AI. So, like, in 1980 I worked on neural networks—and didn't get them to do anything interesting. And every few years after that I would think some more about the computable knowledge problem.

But then I did the science in *A New Kind of Science*—and I discovered this thing I call the Principle of Computational Equivalence, which says many things. But one of them is that there can't be a bright line between the "intelligent" and the "merely computational." So that made me start to think that maybe I didn't need to build a brain to solve the computable knowledge problem.

Meanwhile, my younger son, who I think was about six at the time, was starting to use Mathematica a bit. And he asked me, "Why can't I just tell it what I want to in plain English?" I started explaining how hard that was. But he persisted with, "Well, there just aren't that many different ways to say any particular thing," etc. And that got me thinking—particularly about using the science I'd built to try to solve the problem of understanding natural language.

Meanwhile, I'd started a project to curate lots of data of all kinds. It was an interesting thing going into a big reference library and figuring out what it would take to just make all of that computable. Alan Turing had done some estimates of things like that, which were a bit daunting. But anyway, I started getting all kinds of experts on all kinds of topics that tech companies usually don't care about. And I started building technology and a management system for making data computable.

It was not at all clear this was all going to work, and even a lot of my management team was skeptical. "Another WolframTones" was a common characterization. But the good news was that our main business was strong. And—even though I'd considered it in the early 1990s—I'd never taken the company public, and I didn't have any investors at all, except I guess myself. So I wasn't really answering to anyone. And so I could just do Wolfram|Alpha—as I have been able to do all kinds of long-term stuff throughout the history of our company.

And despite the concerns, Wolfram|Alpha did work. And I have to say that when it was finally ready to demo, it took only one meeting for my management team to completely come around, and be enthusiastic about it.

One problem, of course, with Wolfram|Alpha is that—like Mathematica and the Wolfram Language—it's really an infinite project. But there came a point at which we really couldn't do much more development without seeing what would happen with real users, asking real questions, in real natural language.

So we picked May 15, 2009, as the date to go live. But there was a problem: we had no idea how high the traffic would spike. And back then we couldn't use Amazon or anything: to get performance we had to do fancy parallel computations right on the bare metal.

Michael Dell was kind enough to give us a good deal on getting lots of computers for our colos. But I was pretty concerned when I talked to some people who'd had services that had crashed horribly on launch. So I decided on a kind of hack. I decided that we'd launch on live internet TV—so if something horrible happened, at least people would know what was going on, and might have some fun with it. So I contacted Justin Kan, who was then doing justin.tv, and whose first company I'd failed to invest in at the very first Y Combinator—and we arranged to "launch live."

It was fun building our "mission control"—and we made some very nice dashboards, many of which we actually still use today. But the day of the launch I was concerned that this was going to be the most boring TV ever: that basically at the appointed hour, I'd just click a mouse and we'd be live, and that'd be the end of it.

Well, that was not to be. You know, I've never watched the broadcast. I don't know how much it captures of some of the horrible things that went wrong—particularly with last-minute network configuration issues.

But perhaps the most memorable thing had to do with the weather. We were in Central Illinois. And about an hour before our grand launch, there was a weather report—that a tornado was heading straight for us! You can see the wind speed spike in the Wolfram|Alpha historical weather data:

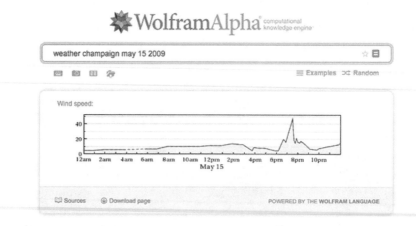

Well, fortunately, the tornado missed. And sure enough, at 9:33:50 pm Central Time on May 15, 2009, I pressed the button, and Wolfram|Alpha went live. Lots of people started using it. Some people even understood that it wasn't a search engine: it was computing things.

The early bug reports then started flowing in. This was the thing Wolfram|Alpha used to do at the very beginning, when something failed:

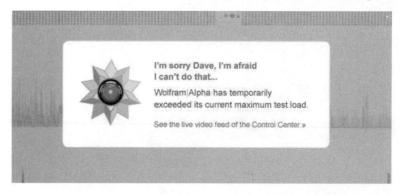

And one of the bug reports was someone saying, "How did you know my name was Dave?!" All kinds of bug reports came in the first night—here are a couple:

> Input: tomato
> Message: What in the world is a tomato's melting point? What
>   happens to a tomato at 31.1 C?
> Romania
> 2009-05-20 05:22:51

> Input: fastest bird
> Message: Ducks are not just food!!! They are faster level flyers than falcons.
>   A frozen chicken will approach 200mph if you drop it from a plane.
> United Kingdom
> 2009-05-20 09:18:11

Well, not only did people start using Wolfram|Alpha; companies did too. Through Bill Gates, Microsoft hooked up Wolfram|Alpha to Bing. And a little company called Siri hooked it up to its app. And some time later Apple bought Siri, and through Steve Jobs, who was by then very sick, Wolfram|Alpha ended up powering the knowledge part of Siri.

OK, so we're getting to modern times. And the big thing now is the Wolfram Language. Actually, it's not such a modern thing for us. Back in the early 1990s I was going to break off the language component of Mathematica—we were thinking of calling it the M Language. And we even had people working on it, like Sergey Brin when he was an intern with us in 1993. But we hadn't quite figured out how to distribute it, or what it should be called.

And in the end, the idea languished. Until we had Wolfram|Alpha, and the cloud existed, and so on. And also I must admit that I was really getting fed up with people thinking of Mathematica as being a "math thing." It had been growing and growing:

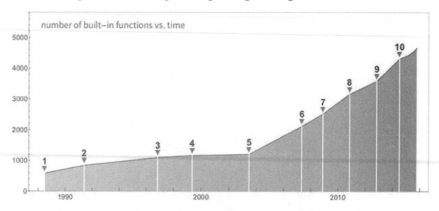

And although we kept on strengthening the math, 90% of it wasn't math at all. We had kind of a "let's just implement everything" approach. And that had gone really well. We were really on a roll inventing all those meta-algorithms, and automating things. And combined with Wolfram|Alpha I realized that what we had was a new, very general kind of thing: a knowledge-based language that built in as much knowledge about computation and about the world as possible.

And there was another piece too: realizing that our symbolic programming paradigm could be used to represent not just computation, but also deployment, particularly in the cloud.

Mathematica has been very widely used in R&D and in education—but with notable exceptions, like in the finance industry, it's not been so widely used for deployed production systems. And one of the ideas of the Wolfram Language—and our cloud—is to change that, and to really make knowledge-based programming something that can be deployed everywhere, from supercomputers to embedded devices. There's a huge amount to say about all this....

And we've done lots of other things too. This shows function growth over the first 10,000 days of Mathematica, what kinds of things were in it over the years.

10,000 Days of Mathematica

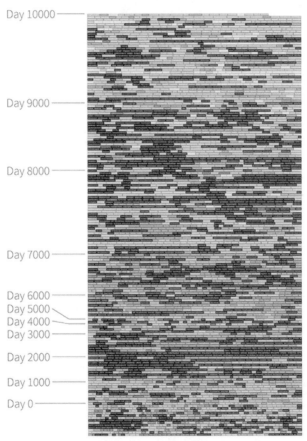

function growth over time (colored by area)

We've done all kinds of different things with our technology. I don't know why I have this picture here, but I have to show it anyway; this was a picture on the commemorative T-shirt for our Image Identification

Project that we did a year ago. Maybe you can figure out what the caption on this means with respect to debugging the image identifier: it was an anteater in the image identifier because we lost the aardvark, who is pictured here:

It was an anteater because we lost the aardvark.

And just in the last few weeks, we've opened up our Wolfram Open Cloud to let anyone use the Wolfram Language on the web. It's really the culmination of 30, perhaps 40, years of work.

You know, for nearly 30 years I've been working hard to make sure the Wolfram Language is well designed—that as it gets bigger and bigger all the pieces fit nicely together, so you can build on them as well as possible. And I have to say it's nice to see how well this has paid off now.

It's pretty cool. We've got a very different kind of language—something that's useful for communicating not just about computation, but about the world, with computers and with humans. You can write tiny programs. There's Tweet-a-Program for example:

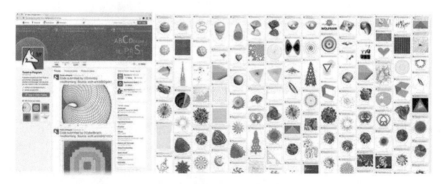

Or you can write big programs—like Wolfram|Alpha, which is 15 million lines of Wolfram Language code.

It's pretty nice to see companies in all sorts of industries starting to base their technology on the Wolfram Language. And another thing I'm really excited about right now is that with the Wolfram Language I think we finally have a great way to teach computational thinking to kids. I even wrote a book about that recently:

And I can't help wondering what would have happened if the 12-year-old me had had this—and if my first computer language had been the Wolfram Language rather than the machine code of the Elliott 903. I could certainly have made some of my favorite science discoveries with one-liners. And a lot of my questions about things like AI would already have been answered.

But actually I'm pretty happy to have been living at the time in history I have, and to have been able to be part of these decades in the evolution of the incredibly important idea of computation—and to have had the privilege of being able to discover and invent a few things relevant to it along the way.

Something I Learned in Kindergarten

May 20, 2016

Fifty years ago today there was a six-year-old at a kindergarten ("nursery school" in British English) in Oxford, England, who was walking under some trees and noticed that the patches of light under the trees didn't look the same as usual. Curious, he looked up at the Sun. It was bright, but he could see that one side of it seemed to be missing. And he realized that was why the patches of light looked odd.

He'd heard of eclipses. He didn't really understand them. But he had the idea that that was what he was seeing. Excited, he told another kid about it. They hadn't heard of eclipses. But he pointed out that the Sun had a bite taken out of it. The other kid looked up. Perhaps the Sun was too bright, but they looked away without noticing anything. Then the first kid tried another kid. And then another. None of them believed him about the eclipse and the bite taken out of the Sun.

Of course, this is a story about me. And now I can find the eclipse by going to Wolfram|Alpha (or the Wolfram Language):

And, yes, it was fun to see my first eclipse (almost exactly 25 years later, I finally saw a total eclipse too). But my real takeaway from that day was about the world and about people. Even if you notice something as obvious as a bite taken out of the side of the Sun, there's no guarantee that you can convince anyone else that it's there.

It's been very helpful to me over the past fifty years to understand that. There've been so many times in my life in science, technology, and business where things seemed as obvious to me as the bite taken out of the Sun. And quite often it's been easy to get other people to see them too. But sometimes they just don't.

When they find out that people don't agree with something that seems obvious to them, many people will just conclude that they're the ones who are wrong. That even though it seems obvious to them, the "crowd" must be right, and they themselves must somehow be confused. Fifty years ago today I learned that wasn't true. Perhaps it made me more obstinate, but I could list quite a few pieces of science and technology that I rather suspect wouldn't exist today if it hadn't been for that kindergarten experience of mine.

As I write this, I feel an urge to tell a few other stories—and lessons learned—from kindergarten. I should explain that I went to a kindergarten with lots of smart kids, mostly children of Oxford academics. They certainly seemed very bright to me at the time—and, interestingly, many of them have ended up having distinguished lives and careers.

In many ways, the kids were much brighter than most of the teachers. I remember one teacher with the curious theory that children's minds were like elastic bands—and that if children learned too much, their minds would snap. Of course, those were the days when Bible Study was part of pretty much any school's curriculum in the UK, and it was probably very annoying that I would come in every day and regale everyone with stories about dinosaurs and geology when the teacher just wanted people to learn Genesis stories.

I don't think I was great at "doing what the other kids do." When I was three years old, and first at school, there was a time when everyone was supposed to run around "like a bus" (I guess ignoring the fact that buses go on roads...). I didn't want to do it, and just stood in one place. "Why aren't you being a bus?" the teacher asked. "Well, I am a lamp post," I said. They seemed sufficiently taken aback by that response that they left me alone.

I learned an important lesson when I was about five, from another kid. We were supposed to be hammering nails into pieces of wood. Yes, in those days in the UK they let five-year-olds do that. Anyway, she had the hammer and said, "Can you hold the nail? Trust me, I know what I'm doing." Needless to say, she missed the nail. My thumb was black for several days. But it was a small price to pay for a terrific life lesson: just because someone claims to know what they're talking about doesn't mean they do. And nowadays, when I'm dealing with some expert who says "trust me, I know what I'm talking about," I can't help but have my mind wander back half a century to that moment just before the hammer fell. The individual involved in this story is now a very distinguished mathematician... presumably using much safer tools.

I'll relate two more stories. The first one I'm not sure how I feel about now. It had to do with learning addition. Now, realistically, I have a good memory (which is perhaps obvious given that I'm writing about things that happened 50 years ago). So I could perfectly well have just memorized all my addition facts. But somehow I didn't want to. And one day I noticed that if I put two rulers next to each other, I could make a little machine that would add for me—an "addition slide rule." So whenever we were doing additions, I always "happened" to have two rulers on my desk. When it came to multiplication, I didn't memorize that either—though in that case I discovered I could go far by knowing the single fact that $7 \times 8 = 56$—because that was the fact other kids didn't know. (In the end, it took until I was in my forties before I'd finally learned every part of my multiplication table up to 12×12.) And as I look at Wolfram|Alpha and Mathematica and so on, and think about my addition slide rule, I'm reminded of the theory that people never really change....

My final story comes from around the same time as the eclipse. Back then, the UK used non-decimal currency: there were 12 pennies in a shilling, and 20 shillings in a pound. And one of the exercises for us kids was to do mixed-radix arithmetic with these things. I was very pleased with myself one day when I figured out that money didn't have to work this way; that everything could be base 10 (well, I didn't explicitly know the concept of base 10 yet). I told this to a teacher. They were a little confused, but said that currency had worked the same way for hundreds of years, and wasn't going to change. A couple of years later, the UK announced it was going to decimalize its currency. (I suspect if it had waited longer it would still have non-decimal currency, and there would just be a big market for calculators that could compute with it.) I've kept this little incident with me all these years—as a reminder that things can change, even if they've been the way they are for a very long time. Oh, and again, that one shouldn't necessarily believe what one's told. But I guess that's a theme....

Music, Mathematica, and the Computational Universe

June 17, 2011

This week I'm giving a talk at a conference on Mathematics and Computation in Music (MCM 2011)... so I decided to collect some of my thoughts on such topics....

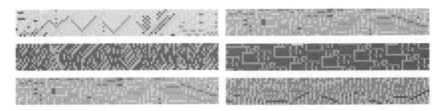

How difficult is it to generate human-like music? To pass the analog of the Turing test for music?

Though music typically has a certain formal structure—as the Pythagoreans noted 2500 years ago—it seems at its core somehow fundamentally human: a reflection of raw creativity that is almost a defining characteristic of human capabilities.

But what is that creativity? Is it something that requires the whole history of our biological and cultural evolution? Or can it exist just as well in systems that have nothing directly to do with humans?

In my work on *A New Kind of Science*, I studied the computational universe of possible programs—and found that even very simple programs can show amazingly rich and complex behavior, on a par, for example, with what one sees in nature. And through my Principle of Computational Equivalence I came to believe that there can be nothing that fundamentally distinguishes our human capabilities from all sorts of processes that occur in nature—or in very simple programs.

But what about music? Some people used their belief that "no simple program will ever create great music" to argue that there must be something wrong with my Principle of Computational Equivalence.

So I became curious: is there really something special and human about music? Or can it in fact be created perfectly well in an automatic, computational way?

In 2003, after my decade as a recluse working on *A New Kind of Science*, I started to be out and about more—and kept on having the mundane problem that my cellphone had the same ringtone as lots of others. So I thought: if distinctive original music could in fact be generated automatically, then one could just "mass customize" cellphone ringtones, and everyone could have their own.

A little while later we decided to try some experiments—and see just what might be possible in creating music from programs.

There's a long history of attempts to produce music from rules. Most of it seems either too robotic or too random. But the discoveries I made in *A New Kind of Science* seemed to offer new possibilities—because they showed that even with the rules of a simple program, it was possible to produce the kind of richness and complexity that, for example, we see and admire in nature.

We started with the most obvious experiment: take the cellular automata that I had studied so much, and use slices of the patterns they generate to form musical scores. I had no real idea what the result of this would be. And certainly some cellular automata with simple patterns of behavior produced completely boring music. But somewhat to my surprise, one really didn't have to go far in the computational universe of possible cellular automata before one started to find remarkably rich and pleasing pieces of music.

The fact that there was always just a simple program underneath gave a certain inevitable logic to the music. But the key point from the science was that even though the underlying program was simple, the pattern it produced could be rich and complex.

But would it be aesthetic? In the visual domain, I had known for a long time that cellular automata could produce pleasing and interesting patterns. And in a sense, given my scientific discoveries, this wasn't surprising. Because I knew that cellular automata could capture the essence of many processes in nature. And insofar as we find nature to be aesthetic, so also this should be true of cellular automata.

But whereas nature just uses a few particular kinds of rules, the complete universe of cellular automata is infinite. In a sense, that computational universe generalizes our actual universe. It keeps the essential mechanisms, but allows an infinite diversity of variations— each with aesthetics that generalize the aesthetics of the natural world.

Ever since the beginning of the 1990s, Mathematica had supported sound generation. And with its symbolic language, Mathematica provided the ideal platform for us to implement our algorithms and start generating music. The results greatly exceeded even our most positive expectations. We used ideas from music theory to take raw cellular automaton creations and "dress them"—and very soon we were producing orchestrated musical pieces that sounded remarkably good.

Around our offices people would sometimes overhear what was being produced—and stop to ask, "What song are you listening to?" We were making music that was good enough that people assumed it must have the usual human origins: we had succeeded in passing the analog of the Turing test for music.

Well, we soon built a website that we called WolframTones. And all sorts of people started using it. I must say that I thought it was an interesting intellectual experiment—and perhaps a good way to make simple ringtones—but not something that one would take very seriously from a musical point of view.

But I was wrong. Pretty quickly all sorts of serious composers started using the site. They would tell us that they found it useful as a source of ideas—as a source of creative seeds for their compositions. In a sense this was bizarre. We had started unsure of whether computers could achieve anything close to human creativity. Yet now skilled humans were coming to our automated system to seek what we might have thought was that uniquely human thing: creative inspiration.

To me, this was a nice validation of the Principle of Computational Equivalence. As one researcher put it, "Once one's heard the music they produce, simple programs seem a lot more like us."

Out in the computational universe, each program in effect defines its own artificial world—whose sounds and logic we get to hear in the music it produces. Some of those worlds are boring, arid places that yield dull, monotonous music. Others are rife with randomness and noise. But every hundred or thousand programs, one finds something wonderful: rich, textured, sometimes familiar, sometimes exotic musical form.

On the WolframTones website we let people press buttons to go and search at random for music that fits into heuristics we've defined for various standard musical genres. We also let people incrementally modify the rules for a piece of music—in effect applying artificial selection to evolve variations they want. And when one uses WolframTones, it feels a bit like doing nature photography. We explore the computational universe to find those corners—those particular programs—that have the significance or aesthetics that we want.

The WolframTones website went live on September 16, 2005. And ever since then it's just been out there on the web, running Mathematica, and creating music. I must admit that I hadn't looked at its logs for quite a while. But doing that now, I discover that it has been used tens of millions of times—creating tens of millions of musical compositions.

By the standards, for example, of Wolfram|Alpha usage, that's nothing. But by the standards of musical composition, it's huge. iTunes now has about 14 million pieces on it—representing most of the published musical output of our species. But in just a few short years, WolframTones has created more compositions than that. By pure computation, it has in a sense surpassed our species in musical output, single-handedly creating more original music than in the whole history of music before it.

To allow instant output, the website encodes music using MIDI (something that the Mathematica language now supports in a direct symbolic way). Many arrangements of WolframTones output as MP3 have been made. And in a peculiar reversal of roles, I went to a recital a few years ago where human performers were playing on violins a piece that had been entirely created using WolframTones methods.

Can simple programs create a complete symphony? A WolframTones composition explores for perhaps a minute the story of some particular computational world. My experience is that to create a longer piece—that tells a bigger story—seems to require higher-level structure. But there is nothing wrong with having a simple program provide that structure. The overarching story it tells can be perfectly compelling, just like so many stories that play out under the aegis of natural laws in the natural world.

But just how much can come from how little? What is the shortest program that makes an interesting musical piece?

It's easy to start constructing Mathematica programs.

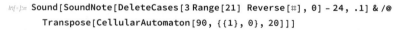

In[◦]:= `Sound[SoundNote[DeleteCases[3 Range[21] Reverse[#], 0] - 24, .1] & /@`
`Transpose[CellularAutomaton[90, {{1}, 0}, 20]]]`

Out[◦]=
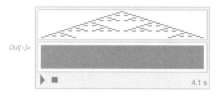

In[◦]:= `Sound[SoundNote[#, 1 / 6, "Warm"] & /@ (Pick[{0, 5, 9, 12, 16, 21}, #, 1] & /@`
`CellularAutomaton[30, {{1, 0, 0, 0, 0, 0}, 0}, 13, {13, 5}])]`

Out[◦]=

And we're planning to do a competition to see how good this can get, especially using all the modern algorithmic tools—like image processing, for example—that exist in Mathematica. But ultimately in such a quest we can't rely on human creativity alone. We have to, in effect, automate this creativity—going beyond what humans have imagined, and instead just exploring the computational universe, and plucking from it the ideas and programs we want.

In creating music we can operate at the level of notes, or collections of notes—or even sound waveforms, generalizing the ways of constructing pleasing waveforms that physical musical devices (or their synthesized direct analogs) have traditionally used.

Of course, creativity from the computational universe is not limited to music. There's been quite a lot of investigation, for example, in the visual arts, and in architecture. Can we create a building from a single, simple rule? If we can, the building will necessarily have a certain logic to its structure, that will allow humans to learn and be comfortable with it.

Can we really appreciate music or other forms that have been created automatically? Or do we always need a story that links what we see into the whole fabric of human culture? Once again, our appreciation of nature makes it clear that no human story is needed. Instead, what seems to be necessary is a connection to a certain overarching logic, which in a sense is precisely what the whole concept of the computational universe provides.

When I look at Wolfram|Alpha, I'm pleased at how much of systematic human knowledge we're being able to capture, and make computable. A new frontier is to capture not just knowledge, but also creativity. To be able, for example, to go from a goal, and creatively work out how to achieve it. Music exposes us to a rather pure form of creativity—and what we have learned, as the Principle of Computational Equivalence might suggest, is that even in this domain, ideas like WolframTones do remarkably well at achieving creative output.

We're going to be able to do another level of automation—in a sense dramatically broadening access to creativity, and no doubt enabling all sorts of fascinating new possibilities.

Ten Thousand Hours of Design Reviews

January 10, 2008

It's not easy to make a big software system that really fits together. It's incredibly important, though. Because it's what makes the whole system more than just the sum of its parts. It's what gives the system limitless possibilities—rather than just a bunch of specific features.

But it's hard to achieve. It requires maintaining consistency and coherence across every area, over the course of many years. But I think it's something we've been very successful at doing with Mathematica. And I think it's actually one of the most crucial assets for the long-term future of Mathematica.

It's also a part of things that I personally am deeply involved in.

Ever since we started developing it more than 21 years ago, I've been the chief architect and chief designer of Mathematica's core function-ality. And particularly for Mathematica 6, there was a huge amount of design to do. Actually, I think much more even than for Mathematica 1.

In fact, I just realized that over the course of the decade during which we were developing Mathematica 6—and accelerating greatly towards the end—I spent altogether about 10,000 hours doing what we call "design reviews," trying to make all those new functions and pieces of functionality in Mathematica 6 be as clean and simple as possible, and all fit together.

At least the way I do it, doing software design is a lot like doing fundamental science.

In fundamental science, one starts from a bunch of phenomena, and then one tries to drill down to find out what's underneath them—to try to find the root causes, the ultimate primitives, of what's going on.

Well, in software design, one starts from a bunch of functionality, and then one needs to drill down to find out just what ultimate primitives one needs to support them.

In science, if one does a good job at finding the primitives, then one can have a very broad theory that covers not just the phenomena one started from, but lots of others too.

And in software design, it's the same kind of thing.

If one does a good job at finding the primitives, then one can build a very broad system that gives one not just the functionality one was first thinking about, but lots more too.

Over the years, we've developed a pretty good process for doing design reviews.

We start with some particular new area of functionality. Then we get a rough description of the functions—or whatever—that we think we'll need to cover it. Then we get down to the hard job of design analysis. Of trying to work out just what the correct fundamental primitives to cover the area are. The clean, simple functions that represent the essence of what's going on—and that fit together with each other, and with the rest of Mathematica, to cover what's needed.

Long ago I used to do design analysis pretty much solo. But nowadays our company is full of talented people who help. The focal point is our Design Analysis group, which works with our experts in particular areas to start the process of refining possible designs.

At some point, though, I always get involved. So that anything that's a core function of Mathematica is always something that I've personally design reviewed.

I sometimes wonder whether it's crazy for me to do this. But I think having one person ultimately review everything is a good way to make sure that there really is coherence and consistency across the system. Of course, when the system is as big as Mathematica, doing all those design reviews to my level of perfection takes a long time—about 10,000 hours, in fact.

Design reviews are usually meetings with somewhere between two and twenty people. (Almost always they're done with web conferencing, not in person.)

The majority of the time, there's a preliminary implementation of whatever it is that we're reviewing. Sometimes the people who are in the design review meeting will say "we think we have this mostly figured out." Sometimes they'll say "we can't see how to set this up; we need your help." Either way, what usually happens is that I start off trying out what's been built, and asking lots and lots of questions about the whole area that's involved.

It's sometimes a little weird. One hour I'll be intensely thinking about the higher mathematics of number theory functions. And the next hour I'll be intensely focused on how we should handle data about cities around the world. Or how we should set up the most general possible interfaces to external control devices.

But although the subject matter is very varied, the principles are at some level the same.

I want to understand things at the most fundamental level—to see what the essential primitives should be. Then I want to make sure those primitives are built so that they fit in as well as possible to the whole existing structure of Mathematica—and so they are as easy as possible for people to understand, and work with.

It's often a very grueling process; progressively polishing things until they are as clean and simple as possible.

Sometimes we'll start a meeting with things looking pretty complicated. A dozen functions that use some strange new construct, and have all sorts of weird arguments and options.

It's usually pretty obvious that we have to do better. But figuring out how is often really hard.

There'll usually be a whole series of incremental ideas. And then a few big shifts—which usually come from getting a clearer understanding of what the true core functionality has to be.

Often we'll be talking quite a bit about precedents elsewhere in Mathematica. Because the more we can make what we're designing now be like something we've done before in Mathematica, the better.

For several reasons. First, because it means we're using approaches that we've tested somewhere else before. Second, because it means that what we're doing now will fit in better to what already exists. And third, because it means that people who are already familiar with other things Mathematica does will have an easier time understanding the new things we're adding.

But some of the most difficult design decisions have to do with when to break away from precedent. When is what we're doing now really different from anything else that we've done before? When is it something sufficiently new—and big—that it makes sense to create some major new structure for it?

At least when we're doing design reviews for Mathematica kernel functions, we always have a very definite final objective for our meetings: we want to actually write the reference documentation—the "function pages"—for what we've been talking about. Because that documentation is what's going to provide the specification for the final implementation—as well as the final definition of the function.

It always works pretty much the same way: I'll be typing at my computer, and everyone else will be watching my screen via screen-sharing. And I'll actually be writing the reference documentation for what each function does. And I'll be asking every sentence or so, "Is that really correct? Is that actually what it should do?" And people will be pointing out this or that problem with what we're saying.

It's a good process, that I think does well at concentrating and capturing what we do in design analysis.

One of the things that happens in design reviews is that we finalize the names for functions.

Naming is a quintessential design review process. It involves drilling down to understand with as much as clarity as possible what a function really does, and is really about. And then finding the perfect word or two that captures the essence of it.

The name has to be something that's familiar enough to people who should be using the function that they'll immediately have an idea of what the function does, but that's general enough that it won't restrict what people will think of doing with the function.

Somehow the very texture of the name also has to communicate something about how broad the function is supposed to be. If it's fairly specialized, it should have a specialized-sounding name. If it's very broad, then it can have a much simpler name—often a much more common English word.

I always have a test for candidate names. If I imagine making up a sentence that explains what the function does, will the proposed name be something that fits into that sentence? Or will one end up always saying that the function with name X does something that is described as Y?

Sometimes it takes us days to come up with the right name for a function. But usually one knows when it's right. It somehow just fits. And one can immediately remember it.

In Mathematica 6, a typical case of function naming was Manipulate. It took quite a while to come up with that name.

We created this great function. But what should it be called? Interface? Activate? Dynamic? Live?

What?

Interface might seem good, because, after all, it creates an interface. But it's a particular kind of interface, not a generic one.

Activate might be good, because it makes things active. But again it's too generic.

Dynamic: again it sounds too general, and also a bit too technical. And anyway we wanted to use that name for something else.

Live... that's a very confusing word. It's even hard to parse when one reads it. Does it say "make it alive," or "here's something that is alive," or what?

Well, after a while one realizes that one has to understand with more clarity just what it is that this great new function is doing.

Yes, it's creating an interface. Yes, it's making things active, dynamic, alive. But really, first and foremost, what it's doing is to provide a way to control something. It's attaching knobs and switches and so on to let one control almost anything.

So what about a word like Control? Again, very hard to understand. Is the thing itself a control? Or is it exerting control?

Handle? Again, too hard to understand.

Harness? A little better. But again, some ambiguity. And definitely too much of a "horse" motif.

Yoke? That one survived for several days. But finally the oxen jokes overwhelmed it.

And then came Manipulate.

At first, it was, "Oh, that's too long a word for such a great and important function."

But in my experience it often "feels right" to have a fairly long word for a function that does so much. Of course there were jokes about it sounding "manipulative."

But as we went on talking about the function, we started just calling it Manipulate among ourselves. And everyone who joined the conversation just knew what it meant. And as we went on developing all its detailed capabilities, it still seemed to fit. It gave the right sense of controlling something, and making something happen.

So that's how Manipulate got its name. It's worked well.

Still, in developing Mathematica 6, we had to name nearly 1000 functions. And each name has to last—just as the names in Mathematica 1 have lasted.

Occasionally it was fairly obvious what a function should be called. Perhaps it had some standard name, say in mathematics or computing, such as Norm or StringSplit.

Perhaps it fit into some existing family of names, like ContourPlot3D.

But most of the time, each name took lots and lots of work to invent. Each one is sort of a minimal expression of a concept that a primitive in Mathematica implements.

Unlike human languages that grow and mutate over time, the Wolfram Language has to be defined once and for all, so that it can be implemented, and so that both the computers and the people who use it can know what everything in it means.

As the Mathematica system has grown, it's in some ways become more and more difficult to do the design. Because every new thing that's added has to fit in with more and more that's already there.

But in some ways it's also become easier. Because there are more precedents to draw on. But most importantly, because we've gotten (and I think I personally have gotten) better and better at doing the design. It's not so much that the quality of the results has changed. It's more that we've gotten faster and faster at solving design problems.

There are problems that come up today that I can solve in a few minutes—yet I remember twenty years ago it taking hours to solve similar problems.

Over the years, there've been quite a few "old chestnuts": design problems that we just couldn't crack. Places where we just couldn't see a clean way to add some particular kind of functionality to Mathematica.

But as we've gotten better and better at design, we've been solving more and more of these. Dynamic interactivity was one big example. And in fact Mathematica 6 has a remarkable number of them solved.

Doing design reviews and nailing down the functional design of Mathematica is a most satisfying intellectual activity. It's incredibly diverse in subject matter. And in a sense always very pure.

It's about a huge range of fundamental ideas—and working out how to fit them all together to create a coherent system that all makes sense.

It's certainly as hard as anything I know about in science. But in many ways it's more creative. One's not trying to decode what exists in the world. One's trying to create something from scratch—to build a world that one can then work within.

I use Mathematica every day. And every day I use countless design ideas that make all the pieces fit smoothly together.

And I realize that, yes, those 10,000 hours of design reviews were worth spending. Even just for me, what we did in them will save me countless hours in being able to do so much more with Mathematica, so much more easily.

And now I'm looking forward to all the design reviews we're starting to do for Mathematica 7, and Mathematica 8....

As of 2019, we've reached Mathematica Version 12 (which is also Wolfram Language Version 12)—with more than 6000 built-in functions.

What Should We Call the Language of Mathematica?

February 12, 2013

At the core of Mathematica is a language. A very powerful symbolic language. Built up with great care over a quarter of a century—and now incorporating a huge swath of knowledge and computation.

Millions and millions of lines of code have been written in this language, for all sorts of purposes. And today—particularly with new large-scale deployment options made possible through the web and the cloud—the language is poised to expand dramatically in usage.

But there's a problem. And it's a problem that—embarrassingly enough—I've been thinking about for more than 20 years. The problem is: what should the language be called?

Usually when I discuss our activities as a company, I talk about progress we've made, or problems we've solved. But today I'm going to make an exception, and talk instead about a problem we haven't solved, but need to solve.

You might say, "How hard can it be to come up with one name?" In my experience, some names are easy to come up with. But others are really, really hard. And this is an example of a really, really hard one. (And perhaps the very length of this chapter communicates some of that difficulty...)

Let's start by talking a little about names in general. There are names like, say, "quark," that are in effect just random words. And that have to get all their meaning "externally," by having it explicitly described. But there are others, like "website" for example, that already give a sense of their meaning just from the words or word roots they contain.

I've named all sorts of things in my time. Science concepts. Technologies. Products. Mathematica functions. I've used different approaches in different cases. In a few cases, I've used "random words" (and have long had a Mathematica-based generator of ones that sound good). But much more often I've tried to start with a familiar word or words that capture the essence of what I'm naming.

And after all, when we're naming things related to our company, we already have a "random" base word: "wolfram." For a while I was a bit squeamish about using it, being that it's my last name. But in recent years it's increasingly been the "lexical glue" that holds together the names of most of the things we're doing.

And so, for example, we have products like Wolfram Finance Platform or Wolfram SystemModeler for professional markets that have that "random" wolfram word, but otherwise try to say more or less directly what they are and what they do.

Wolfram|Alpha is aimed at a much broader audience, and is a more complex case. Because in a short name we need to capture an almost completely new concept. We describe Wolfram|Alpha as a "computational knowledge engine." But how do we shorten that to a name?

I spent a very long time thinking about it, and eventually decided that we couldn't really communicate the concept in the name, and instead we should just communicate some of the sense and character of the system. And that was how we ended up with "alpha": with "alphabet simplicity," a connection to language, a technical character, a tentative software step, and the first, the top. And I'm happy to say the name has worked out very well.

OK. So what about the language that we're trying to name? What should it be called?

Well, I'm pretty sure the word "language" should appear in the name, or at least be able to be tacked onto the name. Because if nothing else, what we've got really is quintessentially a language: a set of constructs that can be strung together to represent an infinite range of meanings.

Our language, though, works in a somewhat different way from ordinary human natural language—most importantly, because it's completely executable: as soon as we express something in the language, that immediately gives us a specification for a unique sequence of computational actions that should be taken.

And in this respect, our language is like a typical computer language. But there is a crucial difference, both practical and philosophical. Typical computer languages (like C or Java or Python) have a small collection of simple built-in operations, and then concentrate on ways to organize those operations to build up programs. But in our language—built right into the language—is a huge amount of computational capability and knowledge.

In a typical computer language, there might be libraries that exist for different kinds of computations. But they're not part of the language, and there's no guarantee they fit together or can be built on. But in our language, the concept from the very beginning has been to build as much as possible in, to have a coherent structure in which as much is automated as possible. And in practice this means that our language has thousands of carefully designed functions and structures that automate a vast range of computations and deliver knowledge in immediately usable ways.

So while in some aspects of its basic mode of operation our language is similar to typical computer languages, its breadth and content is much more reminiscent of human languages—and in a sense it generalizes and deepens both concepts of language.

But OK, what should it be called? Well, I first started thinking about this outrageously long ago—actually in 1990. The software world was different then, and there were different ways we might have deployed the language back then. But despite having put quite a bit of software engineering work into it, we in the end never released it at all. And the single largest reason for that, embarrassingly enough, was that we just couldn't come up with a name for it that we liked.

The "default name" that we used in the development process was the *M Language*, with M presumably short for Mathematica. But I never liked this. It seemed too much like C—a language which I'd used a lot, but whose character and capabilities were utterly different from our language. And particularly given the name "C," M seemed to suggest a language somehow based on "math." Yet even at that time—and to a vastly greater extent today—the language is about much, much more than math. Yes, it can do math really well. But it's broad and deep, and can do an immense range of other algorithmic and computational things—and also an increasing range of things related to built-in knowledge.

One might ask why Mathematica is named as it is. Well, that was a difficult naming process too. The original development name for Mathematica was Omega (and there are still filetype registrations for Mathematica based on that). Then there was a brief moment when it was renamed Polymath. Then Technique. And then there were a whole collection of possibilities.

But finally, at the urging of Steve Jobs, we settled on a name that we had originally rejected for being too long: Mathematica. My original conception of the system—as well as the foundations we built for it— went far beyond math. But math was the first really obvious application area—which is why, when Mathematica was first released, we described it as "a system for doing mathematics by computer."

I've always liked Mathematica as a name. And back in 1988 when Mathematica was launched, it introduced in many ways a new type of name for a computer system, with a certain classical stylishness. In the years since, the name Mathematica has been widely imitated (think Modelica, for example). But it's become clear that for Mathematica itself the name "Mathematica" is in some sense much too narrow— because it gives the idea that all that Mathematica does is math.

For our language we don't want to have the same kind of problem. We want a name that communicates the generality and breadth of the language, and is not tied to one particular application area or type of usage. We want a name that makes sense when the language is used to do tiny pieces of interactive work, or to create giant enterprise applications, and to be used by seasoned software engineers, or by casual script tweakers, or by kids getting their first introduction to programming.

My personal analytics data shows that I've been thinking about the problem of naming our language for 23 years—with episodic bursts of activity. As I mentioned, the original internal name was the *M Language*. More recently the default internal name has been the *Wolfram Language*.

Back in the early 1990s, one of my favorite ideas was *Lingua*—the Latin for language (as well, unfortunately, as tongue), analogous to the Latin character of Mathematica. But Lingua just sounded too weird, and the "gwa" was unpronounceable by too many people whose native languages don't contain that sound. There was some brief enthusiasm for *Express* (think "expression", as well as "express train"), but it died quickly.

There were early suggestions from the MathGroup Mathematica community, like *Principia*, *Harmony*, *Unity*, and *Tongue* (in the latter case, a wag pointed out that bugs could be "slips of the tongue"). One summer intern who worked on the language in 1993 was Sergey Brin (later of Google fame); he suggested the name *Thema*—"the heart of mathematica" ("ma-thema-tica"). My own notes from that time record rather classical-sounding name ideas like *Radix*, *Plurum*, *Practica*, and *Programos*. And in addition to thinking a lot about it myself, I asked linguists, classicists, marketers, and poets—as well as a professional naming expert. But somehow every name either said too little or too much, was too "heavy" or too "light", or for some reason or another just sounded silly. And after more than 20 years, we still don't have a name we like.

But now, with all the new opportunities that exist for it, we just have to release the language—and to do that we have to solve the problem of its name. Which is why I've been thinking hard about it again.

So, what do we want to communicate about the language? First and foremost, as I explained earlier, it's not like other languages. In a sense, it's a new kind of language. It's computational, but it's also got intrinsic content: broad knowledge, structures, and algorithms built in. It's a language that's highly scalable: good for programs ranging from the absolutely tiny to the huge. It's a very general language, useful for a great many different kinds of domains. It's a symbolic language with very clear principles, that can describe arbitrary structures as well as arbitrary data.

It's a fusion of many styles of programming, notably functional and pattern based. It's interactive. And it prides itself on coherence of design, and tries to automate as much as possible of what it does.

At this point, we pretty much have to have "wolfram"—or at least some hint of it—in the name. But it would be nice if there was a good short name or nickname too. We want to communicate that the language is something that we as a company take responsibility for, but also that it will be very widely and often freely available—and not some kind of rare expensive thing.

Alright. So an obvious first question is: how are languages typically named? Well, in Wolfram|Alpha, we have data on more than 16,000 human languages, current and former. And, for example, of the 100 with the most speakers, 13% end in -ese (think Japanese), 11% in -ic (think Arabic), 8% in -ian (think Russian), 5% in -ish (think English), and 3% in -ali (think Bengali). (If one looks at more languages, -ian becomes more common, and -an and -yi start to appear often too.) So should our language be called *Wolframese, Wolframic, Wolframian, Wolframish,* or *Wolframaic*? Or perhaps *Wolfese, Wolfic,* or *Wolfish*? Or *Wolfian* or *Wolfan* or *Wolfatic*, or the exotic *Wolfari* or *Wolfala*? Or a variant like *Wolvese* or *Wolvic*? There are some interesting words here, but to me they all sound a bit too much like obscure tribal languages.

OK. So what about computer languages? Well, there's quite a diversity of names. In rough order of their introduction, some notable languages have been: Fortran, LISP, Algol, COBOL, APL, Simula, SNOBOL, BASIC, PL/I, Logo, Pascal, Forth, C, Smalltalk, Prolog, ML, Scheme, C++, Ada, Erlang, Perl, Haskell, Python, Ruby, Java, JavaScript, PHP, C#, .NET, Clojure, and Go.

So how are these names constructed? Some—particularly earlier ones—are abbreviations, like Fortran ("Formula Translation") and APL ("A Programming Language"). Others are names of people (like Pascal, Ada, and Haskell). Others are named for companies, like Erlang ("Ericsson language") and Go ("Google"). And still others are named in whimsical sequences, like BCPL to B to C ("sea") to shell to

Perl ("pearl") to Ruby—or just plain whimsically, like Python ("Monty Python"). And these naming trends just continue if one looks at less well-known languages.

There are two important points here: first, it seems like computer languages can be called pretty much anything; unlike for most human languages (which are usually derivative on place names), no special linguistic indicator seems to have emerged for computer languages. And second, the names of computer languages only rarely seem immediately to communicate the special features or aspirations of a given language. Sometimes they refer to computer-language history, but often they just seem like quite random words.

So for us, this suggests that perhaps we should just use our existing "random word", and call our language the *Wolfram Language*, or *WL*—or conceivably in short form just *Wolfram*.

Or we could start from our "random word" wolfram, and go more whimsical. One possibility that has generated some enthusiasm internally is *Wolf*. Unfortunately wolves tend to have scary associations— but at least the name Wolf immediately suggests an obvious idea for an icon. And we even already have a possible form for it. Because when we introduced special-character fonts for Mathematica in the mid-1990s, we included a \[Wolf] character (🐺) that was based on a little iconic drawing of mine. Dressing this up could give quite a striking language icon—that could even appear as a single character in a piece of text.

There are variants, like *WolframCode* or *WolframScript*—or *Wolfcode* or *Wolfscript*—but these sound either too obscure or too lightweight. Then there's the somewhat inelegant *WolframLang*, or its shorter forms

WolfLang and *WolfLan*, which sound too much like *Wolfgang*. Then there are names like *WolframX* and *WolfX*, but it's not clear the "X" adds much. Same with *WolframQ* or *WolframL*. There's also *WolframPlus* (*Wolfram+*), *WolframStar* (*Wolfram**), or *WolframDot*. Or *Wolfram1* (when's 2?), *WolframCore* (remember core memory?), or *WolframBase*. There are also Greek-letter suffixes, Wolfram|Alpha-style, like *Wolfram Omega* or *Wolfram Lambda* ("wolf," "ram," and "lamb": too many animals!). Or one could go shorter, like the *W Language*, but that sounds too much like C.

Of course, if one's into "wolf whimsical," there are all kinds of places to go. Wolf backwards is *Flow*, though that hardly seems appropriate for a language so far from simple flowcharts. And then there are names like *Howl* and *Growl* which I can't take too seriously. If one goes into wolf folklore, there are plenty of words and names—but they seem more suited to the Middle Ages than the future.

One can go classical, but the Latin word for wolf is *Lupus*, which is also the name of a disease. And the Greek is *Lukos* [λυκος], which just seems like a random word to modern ears. With different case endings, one gets "differently styled" words. But none of the alternate cases or variants of these words (like *Lupum*, *Lupa*, or *Lukon*) are too promising either—though at least I get to use my knowledge of Latin and Greek from when I was a kid to determine that. (And English forms like *Lupine* are amusing, but don't make it.)

And in the direction of whimsical, there are also words like *Tungsten*, the common English name for element 74, whose symbol W stands for "wolfram," and whose most common ore is wolframite. (And no, it was not discovered by an ancestor of mine.)

How about doing something more scientific? Like searching a space of all possible names, "*NKS* style." For example, one can just try adding all possible single letters to "wolfram," giving such unpromising names as *Wolframa*, *Wolframz*, and *Wolframé*. With two letters, one gets things like *Wolframos*, *Wolframix*, and *WolframUp*. One can try just appending all possible short words, to get things like *WolframHow*, *WolframWay*, and *WolframArt*. And it's a single line of code in our unnamed language (or Mathematica) to find the distribution of, say, what follows "am" in typical English words—yielding ideas like *Wolframsu*, *Wolframity*, or the truly unfortunate *Wolframble*.

But what about going in the other direction, and trying to find word forms that actually relate to what we're trying to communicate about the language? A common way to make up new but suggestive forms is to go back to classical or Indo-European roots, and then try to build novel combinations or variants of these. And of course if we use an actual word form from a language, we at least know that it survived the natural selection of linguistic evolution.

There was a time in the past where one could have taken almost any Latin or Greek root, and expected it to be understood in educated company (as perhaps cyber- was when it was introduced from the Greek [κυβερνητης] for steersman or rudder). But in today's world we pretty much have to limit ourselves to roots which are already at least somewhat familiar from existing words.

And in fact, in the relevant area of "semantic space", "lexical space" is awfully crowded with rather common words. "Language", for example, is *lingua* ("linguistics") or *sermo* ("sermon") in Latin, and *glossa* [γλωσσα] ("glossary") or *phone* [φωνη] ("telephone") in Greek. "Computation" is *computatio* in Latin, and *arithmos* [αριθμος] ("arithmetic") or *logismos* [λογισμος] ("logistics") in Greek. "Knowledge" is *scientia* ("science") or *cognitio* ("cognition") in Latin, and *episteme* [επιστημη] ("epistemology"), *mathesis* [μαθησις] ("mathematics"), or *gnosis* [γνωσις] ("diagnosis") in Greek. "Reasoning" is *ratio* ("rational") in Latin, and *logos* [λογος] ("-ology") in Greek. And so on.

But what can we form from these kinds of roots? I haven't been able to find anything terribly appealing. Typically the names are either ugly, or immediately suggest a meaning that is clearly wrong (like *Wolframology* or *Wolfgloss*).

One can look at other languages, and indeed if you just type "translate *word*" into Wolfram|Alpha (and then press More a few times), you can see translations for as many as a few hundred languages. But typically, beyond Indo-European languages, most of the forms that appear seem random to an English speaker. (Bizarrely, for example, the standard transliteration of the word for "wolf" in Chinese is "lang.")

So where can we go from here? One possible direction is this. We've been trying to find a name by modifying or supplementing the word "wolfram," and expecting that the word "language" will just be added as a suffix. But we need to remember that what we have is really a new kind of language—so perhaps it's the word "language" that we should be thinking of modifying.

But how? There are various prefixes—usually Greek or Latin—that get added, for example, to scientific words to indicate some kind of extension or "beyondness": ana-, alto-, dia-, epi-, exa-, exo-, holo-, hyper-, macro-, mega-, meta-, multi-, neo-, omni-, pan-, pleni-, praeter-, poly-, proto-, super-, uber-, ultra-, and so on. And from these *Wolfram hyperlanguage* (WHL?) is perhaps the nicest possibility—though inevitably it sounds a little "hypey," and is perhaps too reminiscent of hypertext and hyperlinks. (Layering on the Greek and Latin, there's *Hyperlingua* too.)

Wolfram superlanguage, *Wolfram omnilanguage*, and *Wolfram megalanguage* all sound strangely "last century." *Wolfram ultralanguage* and *Wolfram uberlanguage* both seem to be "trying a bit too hard," though *Wolfram Ultra* (without the "language" at all) is a bit better. *Wolfram exolanguage* pleasantly shortens to *Wolfex*, but means the wrong thing (think "exoplanet"). *Wolfram epilanguage* (or just *Wolfram Epi*) does better in terms of meaning (think "epistemology"), but sounds very technical.

A rather frustrating case is *Wolfram metalanguage* (WML). It sounds nice, and in Greek even means more or less the correct thing. But "metalanguage" has already come to have a meaning in English (a language about another language)—and it's not the meaning we want. *Wolfram Meta* might be better, but has the same problem.

So, OK, if we can't make a prefix to the word "language" work, how about just adding a word or phrase between "wolfram" and "language"? Obviously the resulting name is going to be long. But perhaps it'll have a nice abbreviation or shortening.

One immediate idea is *Wolfram Knowledge Language* (*WKL*), but this has the problem of sounding like it might just be a knowledge representation language, not a language that actually incorporates lots of knowledge (as well as algorithms, etc.) More accurate would be *Wolfram Knowledge-Based Language* (*Wolfram KBL*), and perhaps whatever the name, "knowledge-based language" could be used as a description.

Another direction is to insert the word "programming". There's of course *Wolfram Programming Language* (*WPL*). But perhaps better is to start by describing the new kind of programming that our language makes possible—which one might call "hyperprogramming", or conceivably "metaprogramming". ("Macroprogramming" might have been nice, but it's squashed by the old concept of "macros".) And so conceivably one could have *Wolfram Hyperprogramming Language* (*WolframHL, WolframHPL,* or *WHL*) or *Wolfram Metaprogramming Language* (*WML*)—or at least one can use "hyperprogramming language" or "metaprogramming language" as descriptions.

OK, so what's the conclusion? I suppose the most obvious metaconclusion is that getting a name for our language is hard. And the maddening thing is that once we do get a name, my whole 20-year quest will be over incredibly quickly. Perhaps the final name will be one we've already considered, but just weren't thinking about correctly (that's basically what happened with the name Mathematica). Or perhaps some flash of inspiration will lead to a new great name (which is basically what happened with Wolfram|Alpha).

What should the name be? I'm hoping to get feedback on the ideas I've discussed here, as well as to get new suggestions. I must say that as I was writing this essay, I was sort of hoping that in the end it would be a waste, and that by explaining the problem, I would solve it myself. But that hasn't happened. Of course, I'll be thrilled if someone else just outright suggests a great name that we can use. But as I've described, there are many constraints, and what I think is more realistic is for people to suggest frameworks and concepts from which we'll get an idea that will lead to the final name.

I'm very proud of the language we've built over all these years. And I want to make sure that it has a name worthy of it. But once we have a name, we will finally be ready to finish the process of bringing the language to the world—and I'll be very excited to see all the things that makes possible.

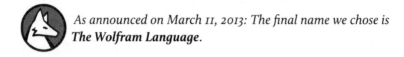   *As announced on March 11, 2013: The final name we chose is* ***The Wolfram Language.***

What Do I Do All Day?
Livestreamed Technology CEOing

December 11, 2017

Thinking in Public

I've been CEOing Wolfram Research for more than 30 years now. But what does that actually entail? What do I end up doing on a typical day? I certainly work hard. But I think I'm not particularly typical of CEOs of tech companies our size. Because for me a large part of my time is spent on the front lines of figuring out how our products should be designed and architected, and what they should do.

Thirty years ago I mostly did this by myself. But nowadays I'm almost always working with groups of people from our 800 or so employees. I like to do things very interactively. And in fact, for the past 15 years or so I've spent much of my time doing what I often call "thinking in public": solving problems and making decisions live in meetings with other people.

I'm often asked how this works, and what actually goes on in our meetings. And recently I realized: what better way to show (and perhaps educate) people than just to livestream lots of our actual meetings? So over the past couple of months, I've livestreamed over 40 hours of my internal meetings—in effect taking everyone behind the scenes in what I do and how our products are created.[*]

---

[*] *As of early 2019, I have livestreamed over 336 hours; see www.stephenwolfram.com/livestreams.*

Seeing Decisions Be Made

In the world at large, people often complain that "nothing happens in meetings." Well, that's not true of my meetings. In fact, I think it's fair to say that in every single product-design meeting I do, significant things are figured out, and at least some significant decisions are made. So far this year, for example, we've added over 250 completely new functions to the Wolfram Language. Each one of those went through a meeting of mine. And quite often the design, the name, or even the very idea of the function was figured out live in the meeting.

There's always a certain intellectual intensity to our meetings. We'll have an hour or whatever, and we'll have to work through what are often complex issues, that require a deep understanding of some area or another—and in the end come up with ideas and decisions that will often have very long-term consequences.

I've worked very hard over the past 30+ years to maintain the unity and coherence of the Wolfram Language. But every day I'm doing meetings where we decide about new things to be added to the language—and it's always a big challenge and a big responsibility to maintain the standards we've set, and to make sure that the decisions we make today will serve us well in the years to come.

It could be about our symbolic framework for neural nets. Or about integrating with databases. Or how to represent complex engineering systems. Or new primitives for functional programming. Or new forms of geo visualization. Or quantum computing. Or programmatic interactions with mail servers. Or the symbolic representation of molecules. Or a zillion other topics that the Wolfram Language covers now, or will cover in the future.

What are the important functions in a particular area? How do they relate to other functions? Do they have the correct names? How can we deal with seemingly incompatible design constraints? Are people going to understand these functions? Oh, and are related graphics or icons as good and clear and elegant as they can be?

By now I basically have four decades of experience in figuring things like this out—and many of the people I work with are also very experienced. Usually a meeting will start with some proposal that's been developed for how something should work. And sometimes it'll just be a question of understanding what's proposed, thinking it through, and then confirming it. But often—in order to maintain the standards we've set—there are real problems that still have to be solved. And a meeting will go back and forth, grappling with some issue or another.

Ideas will come up, often to be shot down. Sometimes it'll feel like we're completely stuck. But everyone in the meeting knows this isn't an exercise; we've got to come up with an actual answer. Sometimes I'll be trying to make analogies—to find somewhere else where we've solved a similar problem before. Or I'll be insisting we go back to first principles—to kind of the center of the problem—to understand everything from the beginning. People will bring up lots of detailed academic or technical knowledge—and I'll usually be trying to extract the essence of what it should be telling us.

It'd certainly be a lot easier if our standards were lower. But we don't want a committee-compromise result. We want actual, correct answers that will stand the test of time. And these often require actual new ideas. But in the end it's typically tremendously satisfying. We put in lots of work and thinking—and eventually we get a solution, and it's a really good solution, that's a real intellectual achievement.

Usually all of this goes on in private, inside our company. But with the livestream, anyone can see it happening—and can see the moment when some function is named, or some problem is solved.

What Are the Meetings Like?

What will actually be going on if you tune in to a livestream? It's pretty diverse. You might see some new Wolfram Language function being tried out (often based on code that's only days or even hours old). You might see a discussion about software engineering, or trends in machine learning, or the philosophy of science, or how to handle some issue of popular culture, or what it's going to take to fix some conceptual bug. You might see some new area get started, you might see some specific piece of Wolfram Language documentation get finished, or you might see a piece of final visual design get done.

There's quite a range of people in our meetings, with a whole diversity of accents and backgrounds and specialties. And it's pretty common for us to need to call in some extra person with specific expertise we hadn't thought was needed. (I find it a little charming that our company culture is such that nobody ever seems surprised to be called into a meeting and asked about a detail of some unusual topic they had no idea was relevant to us before.)

We're a very geographically distributed company (I've been a remote CEO since 1991). So basically all our meetings are through webconferencing. (We use audio and screensharing, but we never find video helpful, except perhaps for looking at a mobile device or a book or a drawing on a piece of paper.)

Most often we're looking at my screen, but sometimes it'll be someone else's screen. (The most common reason to look at someone else's screen is to see something that's only working on their machine so far.) Most often I'll be working in a Wolfram Notebook. Usually there'll be an initial agenda in a notebook, together with executable Wolfram Language code. We'll start from that, but then I'll be modifying the notebook, or creating a new one. Often I'll be trying out design ideas. Sometimes people will be sending code fragments for me to run, or I'll be writing them myself. Sometimes I'll be live-editing our main documentation. Sometimes we'll be watching graphic design being done in real time.

As much as possible, the goal in our meetings is to finish things. To consult in real time with all the people who have input we need, and to get all the ideas and issues about something resolved. Yes, sometimes, afterwards, someone (sometimes me) will realize that something we thought we figured out isn't correct, or won't work. But the good news is that that's pretty rare, probably because the way we run our meetings, things get well aired in real time.

People in our meetings tend to be very direct. If they don't agree with something, they'll say so. I'm very keen that everyone in a meeting actually understands anything that's relevant to them—so we get the benefit of their thinking and judgment about it. (That probably leads to an over-representation from me of phrases like "does that make sense?" or "do you get what I'm saying?")

It really helps, of course, that we have very talented people, who are quick at understanding things. And by now everyone knows that even if the main topic of a meeting is one thing, it's quite likely that we'll have to dip into something completely different in order to make progress. It requires a certain intellectual agility to keep up with this—but if nothing else, I think that's on its own a great thing to practice and cultivate.

For me it's very invigorating to work on so many different topics—often wildly different even between successive hours in a day. It's hard work, but it's also fun. And, yes, there is often humor, particularly in the specifics of the examples we'll end up discussing (lots of elephants and turtles, and strange usage scenarios).

The meetings vary in size from 2 or 3 people to perhaps 20 people. Sometimes people will be added and dropped through the course of the meeting, as the details of what we're discussing change. Particularly in larger meetings—that tend to be about projects that cut across multiple groups—we'll typically have one or more project managers (we call them "PMs") present. The PMs are responsible for the overall flow of the project—and particularly for coordinating between different groups that need to contribute.

If you listen to the livestream, you'll hear a certain amount of jargon. Some of it is pretty typical in the software industry (UX = user experience, SQA = software quality assurance). Some of it is more specific to our company—like acronyms for departments (DQA = Document Quality Assurance, WPE = Web Product Engineering) or names of internal things (XKernel = prototype Wolfram Language build, pods = elements of Wolfram|Alpha output, pinkboxing = indicating undisplayable output, knitting = crosslinking elements of documentation). And occasionally, of course, there's a new piece of jargon, or a new name for something, invented right in the meeting.

Usually our meetings are pretty fast paced. An idea will come up—and immediately people are responding to it. And as soon as something's been decided, people will start building on the decision, and figuring out more. It's remarkably productive, and I think it's a pretty interesting process to watch. Even though without the experience base that the people in the meeting have, there may be some points at which it seems as if ideas are flying around too fast to keep track of what's going on.

The Process of Livestreaming

The idea of livestreaming our internal meetings is new. But over the years I've done a fair amount of livestreaming for other purposes.

Back in 2009, when we launched Wolfram|Alpha, we actually livestreamed the process of making the site live. (I figured that if things went wrong, we might as well just show everyone what actually went wrong, rather than just putting up a "site unavailable" message.)

I've livestreamed demos and explorations of new software we've released. I've livestreamed work I happen to be doing in writing code or producing "computational essays." (My son Christopher is arguably a faster Wolfram Language programmer than me, and he's livestreamed some livecoding he's done too.) I've also livestreamed live experiments, particularly from our Wolfram Summer School and Wolfram Summer Camp.

But until recently, all my livestreaming had basically been solo: it hadn't involved having other people in the livestream. But I've always thought our internal design review meetings are pretty interesting, so I thought "why not let other people listen in on them too?" I have to admit I was a little nervous about this at first. After all, these meetings are pretty central to what our company does, and we can't afford to have them be dragged down by anything.

And so I've insisted that a meeting has to be just the same whether it's livestreamed or not. My only immediate concession to livestreaming is that I give a few sentences of introduction to explain roughly what the meeting is going to be about. And the good news has been that as soon as a meeting gets going, the people in it (including myself) seem to rapidly forget that it's being livestreamed—and just concentrate on the (typically quite intense) things that are going on in the meeting.

But something interesting that happens when we're livestreaming a meeting is that there's real-time text chat with viewers. Often it's questions and general discussion. But sometimes it's interesting comments or suggestions about what we're doing or saying. It's like having instant advisors, or an instant focus group, giving us real-time input or feedback about our decisions.

As a practical matter, the primary people in the meeting are too focused on the meeting itself to be handling text chat. So we have separate people doing that—surfacing a small number of the most relevant comments and suggestions. And this has worked great—and in fact in most meetings at least one or two good ideas come from our viewers, that we're instantly able to incorporate into our thinking.

One can think of livestreaming as something a bit like reality TV— except that it's live and real time. We're planning to have some systematic "broadcast times" for recorded material. But the live component has the constraint that it has to happen when the meetings are actually happening. I tend to have a very full and complex schedule, packing in all the various things I do. And exactly when a particular design review meeting can happen will often depend on when a particular piece of code or design work is ready.

It will also depend on the availability of the various other people in the meetings—who have their own constraints, and often live in a wide range of time zones. I've tried other approaches, but the most common thing now is that design review meetings are scheduled soon before they actually happen, and typically not more than a day or two in advance. And even though I personally work at night as well as during the day, most design reviews tend to get scheduled during US (East Coast) working hours, because that's when it's easiest to arrange for all the people who have to be in the meeting—as well as people who might be called in if their expertise is needed.

From the point of view of livestreaming, it would be nice to have a more predictable schedule of relevant meetings, but the meetings are being set up to achieve maximum productivity in their own right—and livestreaming is just an add-on.

We're trying to use Twitter to give some advance notice of livestreaming. But in the end the best indication of when a livestream is starting is just the notification that comes from the Twitch livestreaming platform we're using. (Yes, Twitch is mainly used for esports right now, but we [and they] hope it can be used for other things too—and with their esports focus, their technology for screensharing has become very good. Curiously, I've been aware of Twitch for a long time. I met its founders at the very first Y Combinator Demo Day in 2005, and we used its precursor, justin.tv, to livestream the Wolfram|Alpha launch.)

Styles of Work

Not all the work I do is suitable for livestreaming. In addition to "thinking in public" in meetings, I also spend time "thinking in private," doing things like just writing. (I actually spent more than 10 years almost exclusively "thinking in private" when I worked on my book *A New Kind of Science*.)

If I look at my calendar for a given week, I'll see a mixture of things. Every day there are typically at least one or two design reviews of the kind I've been livestreaming. There are also a fair number of project reviews, where I'm trying to help move all kinds of projects along. And there are some strategy and management discussions too, along with the very occasional external meeting.

Our company is weighted very heavily towards R&D—and trying to build the best possible products. And that's certainly reflected in the way I spend my time—and in my emphasis on intellectual rather than commercial value. Some people might think that after all these years I couldn't possibly still be involved in the level of detail that's in evidence in the design reviews we've been livestreaming.

But here's the thing: I'm trying hard to design the Wolfram Language in the very best possible way for the long term. And after 40 years of doing software design, I'm pretty experienced at it. So I'm both fairly fast at doing it, and fairly good at not making mistakes. By now, of course, there are many other excellent software designers at our company. But I'm still the person who has the most experience with Wolfram Language design—as well as the most global view of the system (which is part of why in design review meetings, I end up spending some fraction of my time just connecting different related design efforts).

And, yes, I get involved in details. What exactly should the name of that option be? What color should that icon be? What should this function do in a particular corner case? And, yes, every one of these things could be solved in some way without me. But in a fairly short time, I can help make sure that what we have is really something that we can build on—and be proud of—in the years to come. And I consider it a good and worthy way for me to spend my time.

And it's fun to be able to open up this process for people, by livestreaming the meetings we have. I'm hoping it'll be useful for people to understand a bit about what goes into creating the Wolfram Language (and yes, software design often tends to be a bit unsung, and mainly noticed only if it's got wrong—so it's nice to be able to show what's actually involved).

In a sense, doing the design of the Wolfram Language is a very concentrated and high-end example of computational thinking. And I hope that by experiencing it in watching our meetings, people will learn more about how they can do computational thinking themselves.

The meetings that we're livestreaming now are about features of the Wolfram Language etc. that we currently have under development. But with our aggressive schedule of releasing software, it shouldn't be long before the things we're talking about are actually released in working products. And when that happens, there'll be something quite unique about it. Because for the first time ever, people will not only be able to see what got done, but they'll also be able to go back to a recorded livestream and see how it came to be figured out.

It's an interesting and unique record of a powerful form of intellectual activity. But for me it's already nice just to be able to share some of the fascinating conversations I end up being part of every day. And to feel like the time I'm spending as a very hands-on CEO not only advances the Wolfram Language and the other things we're building, but can also directly help educate—and perhaps entertain—a few more people out in the world.

The Story of Spikey

December 28, 2018

Spikeys Everywhere

We call it "Spikey," and in my life today, it's everywhere:

It comes from a 3D object—a polyhedron that's called a rhombic hexecontahedron:

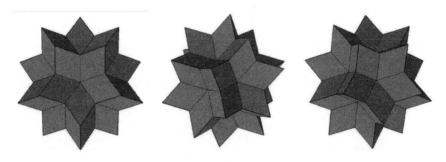

But what is its story, and how did we come to adopt it as our symbol?

The Origins of Spikey

Back in 1987, when we were developing the first version of Mathematica, one of our innovations was being able to generate resolution-independent 3D graphics from symbolic descriptions. In our early demos, this let us create wonderfully crisp images of Platonic solids. But as we approached the release of Mathematica 1.0, we wanted a more impressive example. So we decided to take the last of the Platonic solids—the icosahedron—and then make something more complex by a certain amount of stellation (or, more correctly, cumulation). (Yes, that's what the original notebook interface looked like, 30 years ago....)

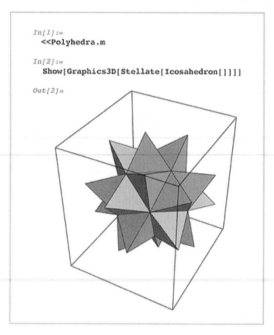

At first this was just a nice demo that happened to run fast enough on the computers we were using back then. But quite soon the 3D object it generated began to emerge as the de facto logo for Mathematica. And by the time Mathematica 1.0 was released in 1988, the stellated icosahedron was everywhere:

In time, tributes to our particular stellation started appearing—in various materials and sizes:

But just a year after we released Mathematica 1.0, we were getting ready to release Mathematica 1.2, and to communicate its greater sophistication, we wanted a more sophisticated logo. One of our developers, Igor Rivin, had done his PhD on polyhedra in hyperbolic space—and through his efforts a hyperbolic icosahedron adorned our Version 1.2 materials:

My staff gave me an up-to-date-Spikey T-shirt for my 30th birthday in 1989, with a quote that I guess even after all these years I'd still say:

After Mathematica 1.2, our marketing materials had a whole collection of hyperbolic Platonic solids, but by the time Version 2.0 arrived in 1991 we'd decided our favorite was the hyperbolic dodecahedron:

Still, we continued to explore other "Spikeyforms." Inspired by the "wood model" style of Leonardo da Vinci's stellated icosahedron drawing (with amazingly good perspective) for Luca Pacioli's book *De divina proportione*, we commissioned a Version 2.0 poster (by Scott Kim) showing five intersecting tetrahedra arranged so that their outermost vertices formed a dodecahedron:

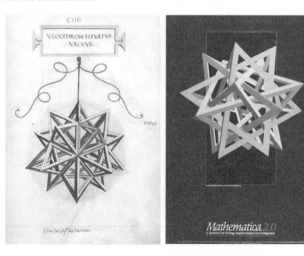

Looking through my 1991 archives today, I find some "explanatory" code (by Ilan Vardi)—and it's nice to see that it all just runs in our latest Wolfram Language (though now it can be written a bit more elegantly):

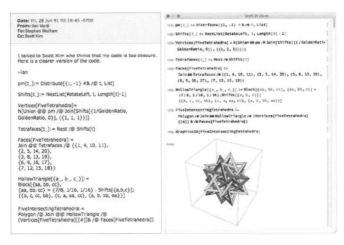

Over the years, it became a strange ritual that when we were getting ready to launch a new integer version of Mathematica, we'd have very earnest meetings to "pick our new Spikey". Sometimes there would be hundreds to choose from, generated (most often by Michael Trott) using all kinds of different algorithms:

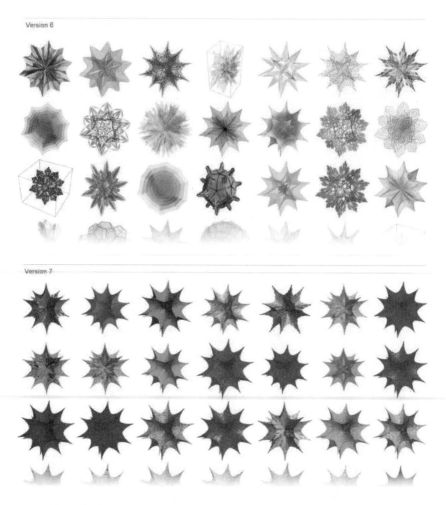

But though the color palettes evolved, and the Spikeys often reflected (though perhaps in some subtle way) new features in the system, we've now had a 30-year tradition of variations on the hyperbolic dodecahedron:

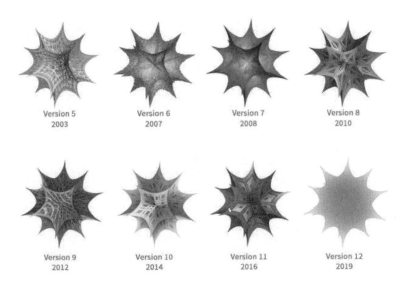

Version 5
2003

Version 6
2007

Version 7
2008

Version 8
2010

Version 9
2012

Version 10
2014

Version 11
2016

Version 12
2019

In more recent times, it's become a bit more streamlined to explore the parameter space—though by now we've accumulated hundreds of parameters:

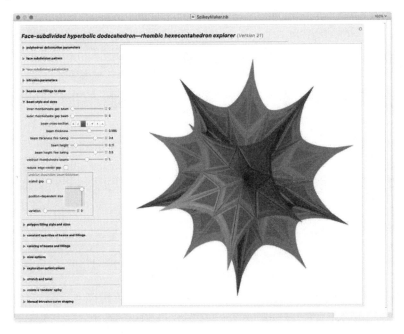

A hyperbolic dodecahedron has 20 points—ideal for celebrating the 20th anniversary of Mathematica in 2008. But when we wanted something similar for the 25th anniversary in 2013 we ran into the problem that there's no regular polyhedron with 25 vertices. But (essentially using SpherePoints[25]) we managed to create an approximate one—and made a 3D printout of it for everyone in our company, sized according to how long they'd been with us:

Enter Wolfram|Alpha

In 2009, we were getting ready to launch Wolfram|Alpha—and it needed a logo. There were all sorts of concepts:

We really wanted to emphasize that Wolfram|Alpha works by doing computation (rather than just, say, searching). And for a while we were keen on indicating this with some kind of gear-like motif. But we also wanted the logo to be reminiscent of our longtime Mathematica logo. So this led to one of those classic "the-CEO-must-be-crazy" projects: make a gear mechanism out of Spikey-like forms.

Longtime Mathematica and Wolfram Language user (and Hungarian mechanical engineer) Sándor Kabai helped out, suggesting a "Spikey Gear":

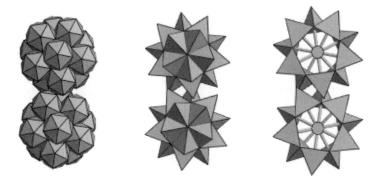

And then, in a throwback to the Version 2 intersecting tetrahedra, he came up with this:

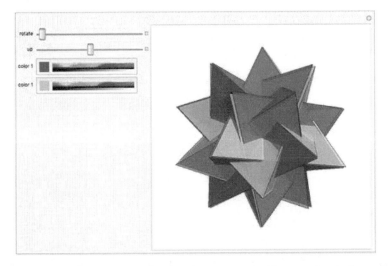

In 2009, 3D printing was becoming very popular, and we thought it would be nice for Wolfram|Alpha to have a logo that was readily 3D printable. Hyperbolic polyhedra were out: their spikes would break off, and could be dangerous. (And something like the Mathematica Version 4 Spikey, with "safety spikes", lacked elegance.)

For a while we fixated on the gears idea. But eventually we decided it'd be worth taking another look at ordinary polyhedra. But if we were going to adopt a polyhedron, which one should it be?

There are of course an infinite number of possible polyhedra. But to make a nice logo, we wanted a symmetrical and somehow "regular" one.

The five Platonic solids—all of whose faces are identical regular polygons—are in effect the "most regular" of all polyhedra:

Then there are the 13 Archimedean solids, all of whose vertices are identical, and whose faces are regular polygons but of more than one kind:

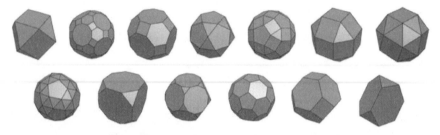

One can come up with all sorts of categories of "regular" polyhedra. One example is the "uniform polyhedra," as depicted in a poster for *The Mathematica Journal* in 1993:

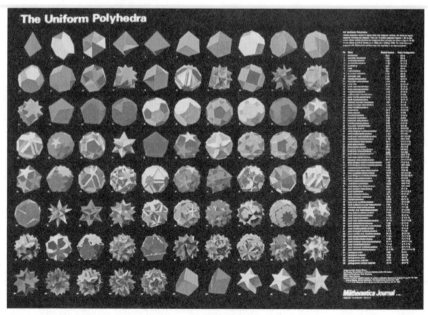

Over the years that Eric Weisstein was assembling what in 1999 became MathWorld, he made an effort to include articles on as many notable polyhedra as possible. And in 2006, as part of putting every kind of systematic data into Mathematica and the Wolfram Language, we started including polyhedron data from MathWorld. The result was that when Version 6.0 was released in 2007, it included the function PolyhedronData that contained extensive data on 187 notable polyhedra:

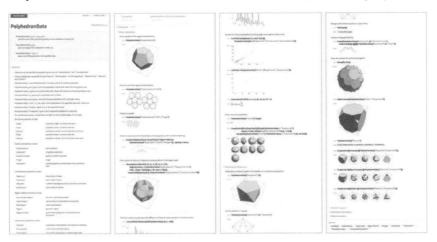

It had always been possible to generate regular polyhedra in Mathematica and the Wolfram Language, but now it became easy. With the release of Version 6.0 we also started the Wolfram Demonstrations Project, which quickly began accumulating all sorts of polyhedron-related Demonstrations.

One created by my then-10-year-old daughter Catherine (who happens to have continued in geometry-related directions) was "Polyhedral Koalas"—featuring a pull-down for all polyhedra in PolyhedronData[]:

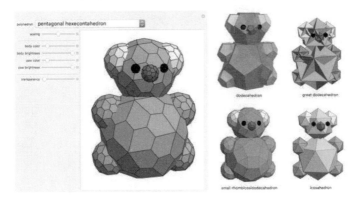

So this was the background when in early 2009 we wanted to "pick a polyhedron" for Wolfram|Alpha. It all came to a head on the evening of Friday, February 6, when I decided to just take a look at things myself.

I still have the notebook I used, and it shows that at first I tried out the rather dubious idea of putting spheres at the vertices of polyhedra:

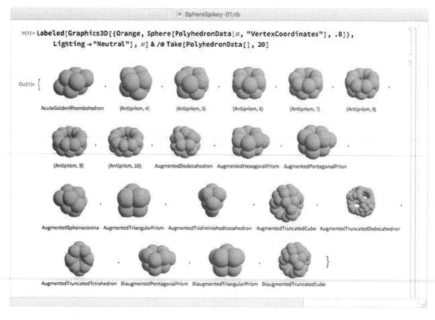

But (as the Notebook History system recorded) just under two minutes later I'd generated pure polyhedron images—all in the orange we thought we were going to use for the logo:

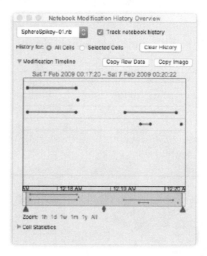

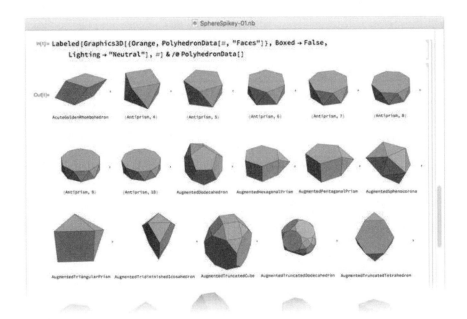

The polyhedra were arranged in alphabetical order by name, and on line 28, there it was—the rhombic hexecontahedron:

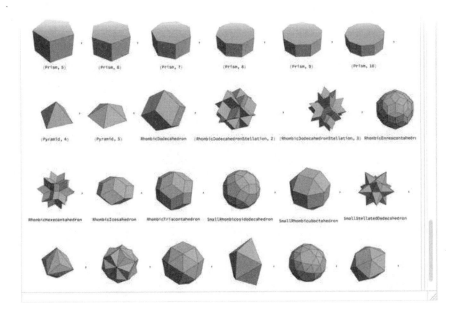

A couple of minutes later, I had homed in on the rhombic hexeconta-hedron, and at exactly 12:24:24 am on February 7, 2009, I rotated it into essentially the symmetrical orientation we now use:

```
In[ ]:= Graphics3D[{Orange, PolyhedronData["RhombicHexecontahedron", "Faces"]},
        Lighting -> "Neutral", Boxed → False]
```

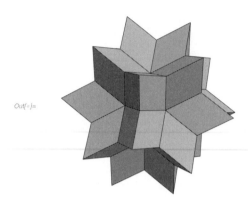

Out[ ]=

I wondered what it would look like in gray scale or in silhouette, and four minutes later I used ColorSeparate to find out:

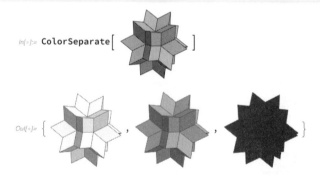

I immediately started writing an email—which I fired off at 12:32 am:

"I [...] rather like the RhombicHexecontahedron ....

It's an interesting shape ... very symmetrical ... I think it might have about the right complexity ... and its silhouette is quite reasonable."

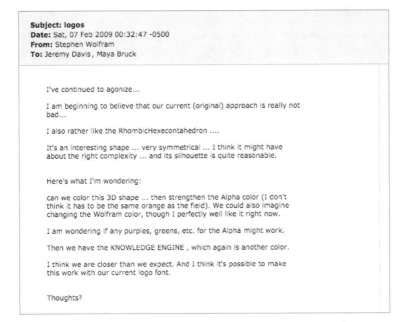

I'd obviously just copied "RhombicHexecontahedron" from the label in the notebook (and I doubt I could have spelled "hexecontahedron" correctly yet). And indeed from my archives I know that this was the very first time I'd ever written the name of what was destined to become my all-time-favorite polyhedron.

It was dead easy in the Wolfram Language to get a picture of a rhombic hexecontahedron to play with:

```
In[·]:= PolyhedronData["RhombicHexecontahedron"]
```

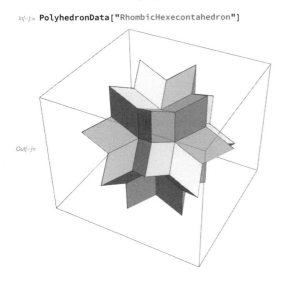

And by Monday it was clear that the rhombic hexecontahedron was a winner—and our art department set about rendering it as the Wolfram|Alpha logo. We tried some different orientations, but soon settled on the symmetrical "head-on" one that I'd picked. (We also had to figure out the best "focal length," giving the best foreshortening.)

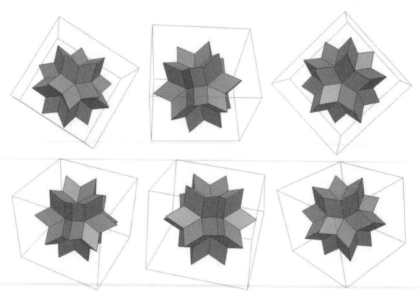

Like our Version 1.0 stellated icosahedron, the rhombic hexecontahedron has 60 faces. But somehow, with its flower-like five-fold "petal" arrangements, it felt much more elegant. It took a fair amount of effort to find the best facet shading in a 2D rendering to reflect the 3D form. But soon we had the first official version of our logo:

It quickly started to show up everywhere, and in a nod to our earlier ideas, it often appeared on a "geared background":

A few years later, we tweaked the facet shading slightly, giving what is still today the logo of Wolfram|Alpha:

The Rhombic Hexecontahedron

What is a rhombic hexecontahedron? It's called a "hexecontahedron" because it has 60 faces, and ἑξήκοντα (hexeconta) is the Greek word for 60. (Yes, the correct spelling is with an "e", not an "a".) It's called "rhombic" because each of its faces is a rhombus. Actually, its faces are golden rhombuses, so named because their diagonals are in the golden ratio $\phi=(1+\sqrt{5})/2 \simeq 1.618$:

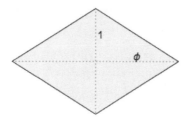

The rhombic hexecontahedron is a curious interpolation between an icosahedron and a dodecahedron (with an icosidodecahedron in the middle). The 12 innermost points of a rhombic hexecontahedron form a regular icosahedron, while the 20 outermost points form a regular dodecahedron. The 30 "middle points" form an icosidodecahedron, which has 32 faces (20 "icosahedron-like" triangular faces, and 12 "dodecahedron-like" pentagonal faces):

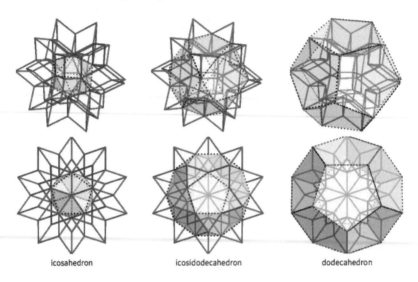

icosahedron   icosidodecahedron   dodecahedron

Altogether, the rhombic hexecontahedron has 62 vertices and 120 edges (as well as 120−62+2=60 faces). There are 3 kinds of vertices ("inner," "middle," and "outer"), corresponding to the 12+30+20 vertices of the icosahedron, icosidodecahedron, and dodecahedron. These types of vertices have respectively 3, 4, and 5 edges meeting at them. Each golden rhombus face of the rhombic hexecontahedron has one "inner" vertex where 5 edges meet, one "outer" vertex where 3 edges meet, and two "middle" vertices where 4 edges meet. The inner and outer vertices are the acute vertices of the golden rhombuses; the middle ones are the obtuse vertices.

The acute vertices of the golden rhombuses have angle $2 \tan^{-1}(\phi^{-1})$ ≈63.43°, and the obtuse ones $2 \tan^{-1}(\phi)$≈116.57°. The angles allow the rhombic hexecontahedron to be assembled from Zometool using only red struts (the same as for a dodecahedron):

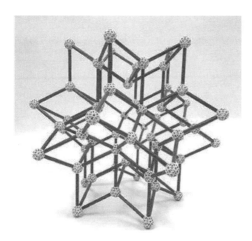

Across the 120 edges of the rhombic hexecontahedron, the 60 "inward-facing hinges" have dihedral angle $4\pi/5=144°$, and the 60 "outward-facing" ones have dihedral angle $2\pi/5=72°$. The solid angles subtended by the inner and outer vertices are $\pi/5$ and $3\pi/5$.

To actually draw a rhombic hexecontahedron, one needs to know 3D coordinates for its vertices. A convenient way to get these is to use the fact that the rhombic hexecontahedron is invariant under the icosahedral group, so that one can start with a single golden rhombus and just apply the 60 matrices that form a 3D representation of the icosahedral group. This gives for example final vertex coordinates $\{\pm\phi, \pm 1, 0\}$, $\{\pm 1, \pm\phi, \pm(1+\phi)\}$, $\{\pm 2\phi, 0, 0\}$, $\{\pm\phi, \pm(1+2\phi), 0\}$, $\{\pm(1+\phi), \pm(1+\phi), \pm(1+\phi)\}$, and cyclic permutations of these, with each possible sign being taken.

In addition to having faces that are golden rhombuses, the rhombic hexecontahedron can be constructed out of 20 golden rhombohedra (whose 6 faces are all golden rhombuses):

There are other ways to build rhombic hexecontahedra out of other polyhedra. Five intersecting cubes can do it, as can 182 dodecahedra with touching faces:

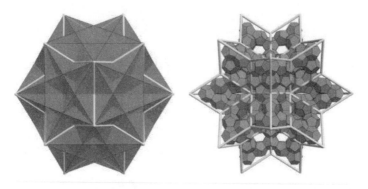

Rhombic hexecontahedra don't tessellate space. But they do interlock in a satisfying way (and, yes, I've seen tens of paper ones stacked up this way):

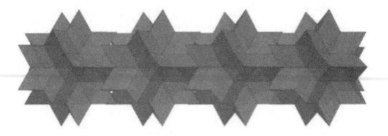

There are also all sorts of ring and other configurations that can be made with them:

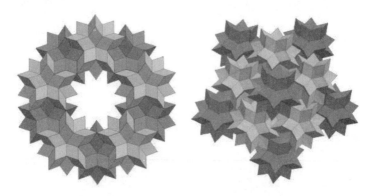

Closely related to the rhombic hexecontahedron ("RH") is the rhombic triacontahedron ("RT"). Both the RH and the RT have faces that are golden rhombuses. But the RH has 60, while the RT has 30. Here's what a single RT looks like:

RTs fit beautifully into the "pockets" in RHs, leading to forms like this:

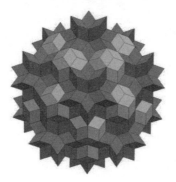

The aforementioned Sándor Kabai got enthusiastic about the RH and RT around 2002. And after the Wolfram Demonstrations Project was started, he and Slovenian mathematician Izidor Hafner ended up contributing over a hundred Demonstrations about RH, RT, and their many properties:

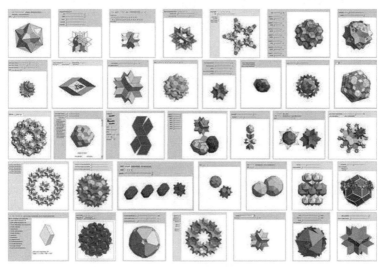

Paper Spikey Kits

As soon as we'd settled on a rhombic hexecontahedron Spikey, we started making 3D printouts of it. (It's now very straightforward to do this with Printout3D[PolyhedronData[...]], and there are also precomputed models available at outside services.)

At our Wolfram|Alpha launch event in May 2009, we had lots of 3D Spikeys to throw around:

But as we prepared for the first post-Wolfram|Alpha holiday season, we wanted to give everyone a way to make their own 3D Spikey. At first we explored using sets of 20 plastic-covered golden rhombohedral magnets. But they were expensive, and had a habit of not sticking together well enough at "Spikey scale."

So that led us to the idea of making a Spikey out of paper, or thin cardboard. Our first thought was then to create a net that could be folded up to make a Spikey:

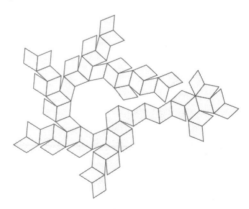

My daughter Catherine was our test folder (and still has the object that was created), but it was clear that there were a lot of awkward hard-to-get-there-from-here situations during the folding process. There are a huge number of possible nets (there are already 43,380 even for the dodecahedron and icosahedron)—and we thought that perhaps one could be found that would work better:

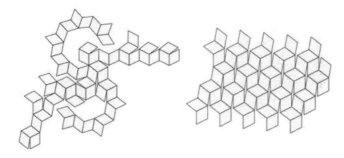

But after failing to find any such net, we then had a new (if obvious) idea: since the final structure would be held together by tabs anyway, why not just make it out of multiple pieces? We quickly realized that the pieces could be 12 identical copies of this:

And with this we were able to create our "Paper Sculpture Kits":

Making the instructions easy to understand was an interesting challenge, but after a few iterations they're now well debugged, and easy for anyone to follow:

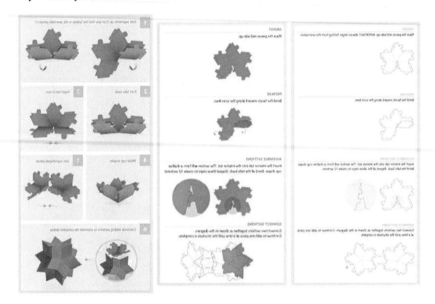

And with paper Spikeys in circulation, our users started sending us all sorts of pictures of Spikeys "on location":

The Path to the Rhombic Hexecontahedron

It's not clear who first identified the Platonic solids. Perhaps it was the Pythagoreans (particularly living near so many polyhedrally shaped pyrite crystals). Perhaps it was someone long before them. Or perhaps it was a contemporary of Plato's named Theaetetus. But in any case, by the time of Plato (≈400 BC), it was known that there are five Platonic solids. And when Euclid wrote his *Elements* (around 300 BC) perhaps the pinnacle of it was the proof that these five are all there can be. (This proof is notably the one that takes the most steps—32—from the original axioms of the *Elements*.)

Platonic solids were used for dice and ornaments. But they were also given a central role in thinking about nature, with Plato for example suggesting that perhaps everything could in some sense be made of them: earth of cubes, air of octahedra, water of icosahedra, fire of tetrahedra, and the heavens ("ether") of dodecahedra.

But what about other polyhedra? In the 4th century AD, Pappus wrote that a couple of centuries earlier, Archimedes had discovered 13 other "regular polyhedra"—presumably what are now called the Archimedean solids—though the details were lost. And for a thousand years little more seems to have been done with polyhedra. But in the 1400s, with the Renaissance starting up, polyhedra were suddenly in vogue again. People like Leonardo da Vinci and Albrecht Dürer routinely used them in art and design, rediscovering some of the Archimedean solids—as well as finding some entirely new polyhedra, like the icosidodecahedron.

But the biggest step forward for polyhedra came with Johannes Kepler at the beginning of the 1600s. It all started with an elegant, if utterly wrong, theory. Theologically convinced that the universe must be constructed with mathematical perfection, Kepler suggested that the six planets known at the time might move on nested spheres geometrically arranged so as to just fit the suitably ordered five Platonic solids between them:

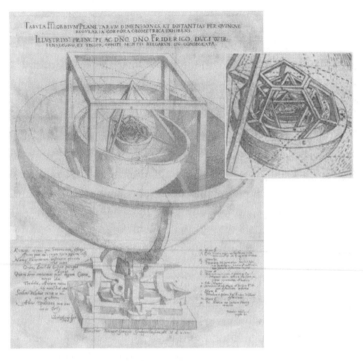

In his 1619 book *Harmonices mundi* ("Harmony of the World") Kepler argued that many features of music, planets, and souls operate according to similar geometric ratios and principles. And to provide raw material for his arguments, Kepler studied polygons and polyhedra, being particularly interested in finding objects that somehow formed complete sets, like the Platonic solids.

He studied possible "sociable polygons," that together could tile the plane—finding, for example, his "monster tiling" (with pentagons, pentagrams, and decagons). He studied "star polyhedra" and found various stellations of the Platonic solids (and in effect the Kepler–Poinsot polyhedra). In 1611 he had published a small book about the hexagonal structure of snowflakes, written as a New Year's gift for a sometime patron of his. And in this book he discussed 3D packings of spheres (and spherical atoms), suggesting that what's now called the Kepler packing (and routinely seen in the packing of fruit in grocery stores) is the densest possible packing (a fact that wasn't formally proved until into the 2000s—as it happens, with the help of Mathematica).

There are polyhedra lurking in Kepler's various packings. Start from any sphere, then look at its neighbors, and join their centers to make the vertices of a polyhedron. For Kepler's densest packing, there are 12 spheres touching any given sphere, and the polyhedron one gets is the cuboctahedron, with 12 vertices and 14 faces. But Kepler also discussed another packing, 8% less dense, in which 8 spheres touch a given sphere, and 6 are close to doing so. Joining the centers of these spheres gives a polyhedron called the rhombic dodecahedron, with 14 vertices and 12 faces:

Having discovered this, Kepler started looking for other "rhombic polyhedra." The rhombic dodecahedron he found has rhombuses composed of pairs of equilateral triangles. But by 1619 Kepler had also looked at golden rhombuses—and had found the rhombic triacontahedron, and drew a nice picture of it in his book, right next to the rhombic dodecahedron:

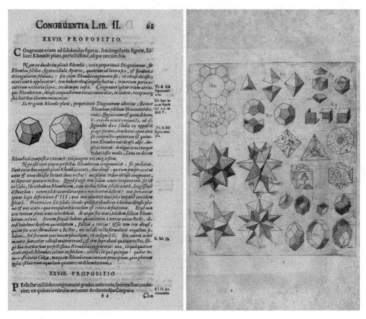

Kepler actually had an immediate application for these rhombic polyhedra: he wanted to use them, along with the cube, to make a nested-spheres model that would fit the orbital periods of the four moons of Jupiter that Galileo had discovered in 1610.

Why didn't Kepler discover the rhombic hexecontahedron? I think he was quite close. He looked at non-convex "star" polyhedra. He looked at rhombic polyhedra. But I guess for his astronomical theories he was satisfied with the rhombic triacontahedron, and looked no further.

In the end, of course, it was Kepler's laws—which have nothing to do with polyhedra—that were his main surviving contribution to astronomy. But Kepler's work on polyhedra—albeit done in the service of a misguided physical theory—stands as a timeless contribution to mathematics.

Over the next three centuries, more polyhedra, with various forms of regularity, were gradually found—and by the early 1900s there were many known to mathematicians:

But, so far as I can tell, the rhombic hexecontahedron was not among them. And instead its discovery had to await the work of a certain Helmut Unkelbach. Born in 1910, he got a PhD in math at the University of Munich in 1937 (after initially studying physics). He wrote several papers about conformal mapping, and—perhaps through studying mappings of polyhedral domains—was led in 1940 to publish a paper (in German) about "The Edge-Symmetric Polyhedra."

His goal, he explains, is to exhaustively study all possible polyhedra that satisfy a specific, though new, definition of regularity: that their edges are all the same length, and these edges all lie in some symmetry plane of the polyhedron. The main result of his paper is a table containing 20 distinct polyhedra with that property:

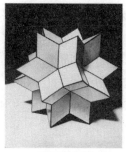

Most of these polyhedra Unkelbach knew to already be known. But Unkelbach singles out three types that he thinks are new: two *hexakisoctahedra* (or disdyakis dodecahedra), two *hexakisicosahedra* (or dysdyakis triacontahedra), and what he calls the *Rhombenhexekontaeder*, or in English, the rhombic hexecontahedron. He clearly considers the rhombic hexecontahedron his prize specimen, including a photograph of a model he made of it:

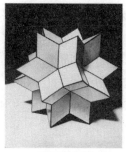

How did he actually "derive" the rhombic hexecontahedron? Basically, he started from a dodecahedron, and identified its two types of symmetry planes:

Then he subdivided each face of the dodecahedron:

Then he essentially considered pushing the centers of each face in or out to a specified multiple α of their usual distance from the center of the dodecahedron:

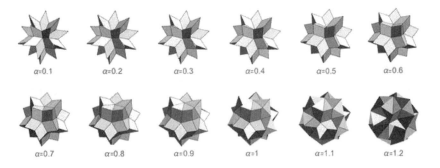

For $\alpha < 1$, the resulting faces don't intersect. But for most values of α, they don't have equal-length sides. That only happens for the specific case $\alpha = 5 - 2\sqrt{5} \approx 0.53$—and in that case the resulting polyhedron is exactly the rhombic hexecontahedron.

Unkelbach actually viewed his 1940 paper as a kind of warmup for a study of more general "k-symmetric polyhedra" with looser symmetry requirements. But it was already remarkable enough that a mathematics journal was being published at all in Germany after the beginning of World War II, and soon after the paper, Unkelbach was pulled into the war effort, spending the next few years designing acoustic-homing torpedoes for the German navy.

Unkelbach never published on polyhedra again, and died in 1968. After the war he returned to conformal mapping, but also started publishing on the idea that mathematical voting theory was the key to setting up a well-functioning democracy, and that mathematicians had a responsibility to make sure it was used.

But even though the rhombic hexecontahedron appeared in Unkelbach's 1940 paper, it might well have languished there forever, were it not for the fact that in 1946 a certain H. S. M. ("Donald") Coxeter wrote a short review of the paper for the (fairly new) American *Mathematical Reviews*. His review catalogs the polyhedra mentioned in the paper, much as a naturalist might catalog new species seen on an expedition. The high point is what he describes as "a remarkable rhombic hexecontahedron," for which he reports that "its faces have the same shape as those of the triacontahedron, of which it is actually a stellation."

Polyhedra were not exactly a hot topic in the mathematics of the mid-1900s, but Coxeter was their leading proponent—and was connected in one way or another to pretty much everyone who was working on them. In 1948 he published his book *Regular Polytopes*. It describes in a systematic way a variety of families of regular polyhedra, in particular showing the great stellated triacontahedron (or great rhombic triacontahedron)—which effectively contains a rhombic hexecontahedron:

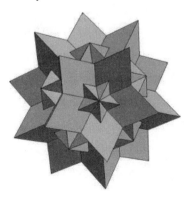

But Coxeter didn't explicitly mention the rhombic hexecontahedron in his book, and while it picked up a few mentions from polyhedron aficionados, the rhombic hexecontahedron remained a basically obscure (and sometimes misspelled) polyhedron.

Quasicrystals

Crystals had always provided important examples of polyhedra. But by the 1800s, with atomic theory increasingly established, there began to be serious investigation of crystallography, and of how atoms are arranged in crystals. Polyhedra made a frequent appearance, in particular in representing the geometries of repeating blocks of atoms ("unit cells") in crystals.

By 1850 it was known that there were basically only 14 possible such geometries; among them is one based on the rhombic dodecahedron. A notable feature of these geometries is that they all have specific two-, three-, four-, or six-fold symmetries—essentially a consequence of the fact that only certain polyhedra can tessellate space, much as in 2D the only regular polygons that can tile the plane are squares, triangles, and hexagons.

But what about for non-crystalline materials, like liquids or glasses? People had wondered since before the 1930s whether at least approximate five-fold symmetries could exist there. You can't tessellate space with regular icosahedra (which have five-fold symmetry), but maybe you could at least have icosahedral regions with little gaps in between.

None of this was settled when in the early 1980s electron diffraction crystallography on a rapidly cooled aluminum-manganese material effectively showed five-fold symmetry. There were already theories about how this could be achieved, and within a few years there were also electron microscope pictures of grains that were shaped like rhombic triacontahedra:

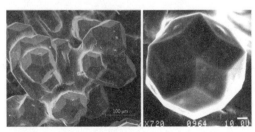

And as people imagined how these triacontahedra could pack together, the rhombic hexecontahedron soon made its appearance—as a "hole" in a cluster of 12 rhombic triacontahedra:

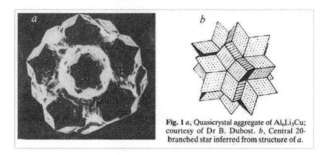

Fig. 1 *a*, Quasicrystal aggregate of Al$_6$Li$_3$Cu; courtesy of Dr B. Dubost. *b*, Central 20-branched star inferred from structure of *a*.

At first it was referred to as a "20-branched star". But soon the connection with the polyhedron literature was made, and it was identified as a rhombic hexecontahedron.

Meanwhile, the whole idea of making things out of rhombic elements was gaining attention. Michael Longuet-Higgins, longtime oceanographer and expert on how wind makes water waves, jumped on the bandwagon, in 1987 filing a patent for a toy based on magnetic rhombohedral blocks, that could make a "Kepler Star" (rhombic hexecontahedron) or a "Kepler Ball" (rhombic triacontahedron):

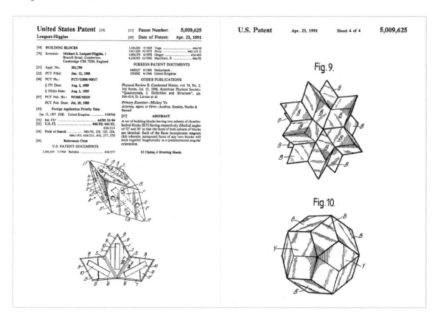

And—although I only just found this out—the rhombohedral blocks that we considered in 2009 for widespread "Spikey making" were actually produced by Dextro Mathematical Toys (aka Rhombo.com), operating out of Longuet-Higgins's house in San Diego.

The whole question of what can successfully tessellate space—or even tile the plane—is a complicated one. In fact, the general problem of whether a particular set of shapes can be arranged to tile the plane has been known since the early 1960s to be formally undecidable. (One might verify that 1000 of these shapes can fit together, but it can take arbitrarily more computational effort to figure out the answer for more and more of the shapes.)

People like Kepler presumably assumed if a set of shapes was going to tile the plane, they must be able to do so in a purely repetitive pattern. But following the realization that the general tiling problem is undecidable, Roger Penrose in 1974 came up with two shapes that could successfully tile the plane, but not in a repetitive way. By 1976 Penrose (as well as Robert Ammann) had come up with a slightly simpler version:

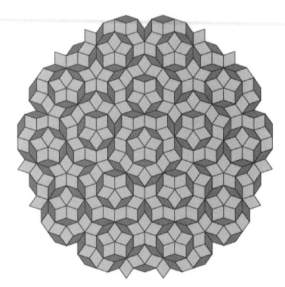

And, yes, the shapes here are rhombuses, though not golden rhombuses. But with angles 36°,144° and 72°,108°, they arrange with 5- and 10-fold symmetry.

By construction, these rhombuses (or, more strictly, shapes made from them) can't form a repetitive pattern. But it turns out they can form a pattern that can be built up in a systematic, nested way:

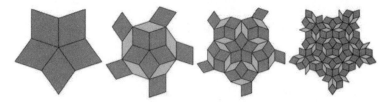

And, yes, the middle of step 3 in this sequence looks rather like our flattened Spikey. But it's not exactly right; the aspect ratios of the outer rhombuses are off.

But actually, there is still a close connection. Instead of operating in the plane, imagine starting from half a rhombic triacontahedron, made from golden rhombuses in 3D:

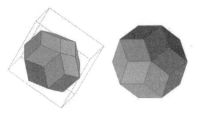

Looking at it from above, it looks exactly like the beginning of the nested construction of the Penrose tiling. If one keeps going, one gets the Penrose tiling:

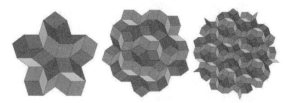

Looked at "from the side" in 3D, one can tell it's still just identical golden rhombuses:

Putting four of these "Wieringa roofs" together one can form exactly the rhombic hexecontahedron:

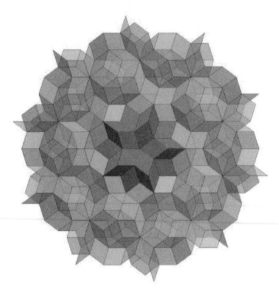

But what's the relation between these nested constructions and the actual way physical quasicrystals form? It's not yet clear. But it's still neat to see even hints of rhombic hexecontahedra showing up in nature.

And historically it was through their discussion in quasicrystals that Sándor Kabai came to start studying rhombic hexecontahedra with Mathematica, which in turn led Eric Weisstein to find out about them, which in turn led them to be in Mathematica and the Wolfram Language, which in turn led me to pick one for our logo. And in recognition of this, we print the nestedly constructed Penrose tiling on the inside of our paper Spikey:

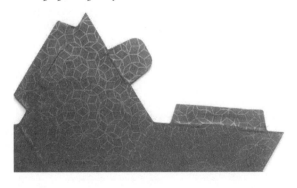

Flattening Spikey

Our Wolfram|Alpha Spikey burst onto the scene in 2009 with the release of Wolfram|Alpha. But we still had our long-running and progressively evolving Mathematica Spikey too. So when we built a new European headquarters in 2011 we had not just one, but two Spikeys vying to be on it.

Our longtime art director Jeremy Davis came up with a solution: take one Spikey, but "idealize" it, using just its "skeleton." It wasn't hard to decide to start from the rhombic hexecontahedron. But then we flattened it (with the best ratios, of course)—and finally ended up with the first implementation of our now-familiar logo:

The Brazilian Surprise

When I started writing this piece, I thought the story would basically end here. After all, I've now described how we picked the rhombic hexecontahedron, and how mathematicians came up with it in the first place. But before finishing the piece, I thought, "I'd better look through all the correspondence I've received about Spikey over the years, just to make sure I'm not missing anything."

And that's when I noticed an email from June 2009, from an artist in Brazil named Yolanda Cipriano. She said she'd seen an article about Wolfram|Alpha in a Brazilian news magazine—and had noticed the

Spikey—and wanted to point me to her website. It was now more than nine years later, but I followed the link anyway, and was amazed to find this:

Enfeite artesanal brasileiro, feito com retalhos de chitas aplicados sobre estrutura geométrica de papel.

História

Imagens

Eu quero!

Contato

Um símbolo feito de estrelas ligadas entre si e gira...

Tem diversos lados, cada um de uma cor É popular feito de retalhos...

Giramundo :: Flor de Mandacarú :: Brasil :: giramundos@gmail.com

I read more of her email: "Here in Brazil this object is called 'Giramundo' or 'Flor Mandacarú' (Mandacaru Flower) and it is an artistic ornament made with [tissue paper]."

What?! There was a Spikey tradition in Brazil, and all these years we'd never heard about it? I soon found other pictures on the web. Only a few of the Spikeys were made with paper; most were fabric—but there were lots of them:

I emailed a Brazilian friend who'd worked on the original development of Wolfram|Alpha. He quickly responded, "These are indeed familiar objects… and to my shame I was never inquisitive enough to connect the dots"—then sent me pictures from a local arts and crafts catalog:

But now the hunt was on: what were these things, and where had they come from? Someone at our company volunteered that actually her great-grandmother in Chile had made such things out of crochet—and always with a tail. We started contacting people who had put up pictures of "folk Spikeys" on the web. Quite often all they knew was that they got theirs from a thrift shop. But sometimes people would say that they knew how to make them. And the story always seemed to be the same: they'd learned how to do it from their grandmothers.

The typical way to build a folk Spikey—at least in modern times—seems to be to start off by cutting out 60 cardboard rhombuses. The next step is to wrap each rhombus in fabric—and finally to stitch them all together:

OK, but there's an immediate math issue here. Are these people really correctly measuring out 63° golden rhombuses? The answer is typically no. Instead, they're making 60° rhombuses out of pairs of equilateral triangles—just like the standard diamond shapes used in quilts. So how then does the Spikey fit together? Well, 60° is not far from 63°, and if you're sewing the faces together, there's enough wiggle room that it's easy to make the polyhedron close even without the angles being precisely right. (There are also "quasi-Spikeys" that—as in Unkelbach's construction—don't have rhombuses for faces, but instead have pointier "outside triangles".)

Folk Spikeys on the web are labeled in all sorts of ways. The most common is as "Giramundos." But quite often they are called "Estrelas da Felicidade" ("stars of happiness"). Confusingly, some of them are also labeled "Moravian stars"—but actually, Moravian stars are different and much pointier polyhedra (most often heavily augmented rhombicuboctahedra) that happen to have recently become popular, particularly for light fixtures.

Despite quite a bit of investigation, I still don't know what the full history of the "folk Spikey" is. But here's what I've found out so far. First, at least what survives of the folk Spikey tradition is centered around Brazil (even though we have a few stories of other appearances). Second, the tradition seems to be fairly old, definitely dating from well before 1900 and quite possibly several centuries earlier. So far as I can tell—as is common with folk art—it's a purely oral tradition, and so far I haven't found any real historical documentation about it.

My best information has come from a certain Paula Guerra, who sold folk Spikeys at a tourist-oriented cafe she operated a decade ago in the historic town of São Luíz do Paraitinga. She said people would come into her cafe from all over Brazil, see the folk Spikeys, and say, "I haven't seen one of those in 50 years…"

Paula herself learned about folk Spikeys (she calls them "stars") from an older woman living on a multigenerational local family farm, who'd been making them since she was a little girl, and had been taught how to do it by her mother. Her procedure—which seems to have been typical—was to get cardboard from anywhere (originally, things like hat boxes), then to cover it with fabric scraps, usually from clothes, then to sew the whole perhaps-6"-across object together.

How old is the folk Spikey? Well, we only have oral tradition to go by. But we've tracked down several people who saw folk Spikeys being made by relatives who were born around 1900. Paula said that a decade ago she'd met an 80-year-old woman who told her that when she was growing up on a 200-year-old coffee farm there was a shelf of folk Spikeys from four generations of women.

At least part of the folk Spikey story seems to center around a mother-daughter tradition. Mothers, it is said, often made folk Spikeys as wedding presents when their daughters went off to get married. Typically the Spikeys were made from scraps of clothes and other things that would remind the daughters of their childhood—a bit like how quilts are sometimes made for modern kids going to college.

But for folk Spikeys there was apparently another twist: it was common that before a Spikey was sewn up, a mother would put money inside it, for her daughter's use in an emergency. The daughter would then keep her Spikey with her sewing supplies, where her husband would be unlikely to pick it up. (Some Spikeys seem to have been used as pincushions— perhaps providing an additional disincentive for them to be picked up.)

What kinds of families had the folk Spikey tradition? Starting around 1750 there were many coffee and sugar plantations in rural Brazil, far from towns. And until perhaps 1900 it was common for farmers from these plantations to get brides—often as young as 13—from distant towns. And perhaps these brides—who were typically from well-off families of Portuguese descent, and were often comparatively well educated—came with folk Spikeys.

In time the tradition seems to have spread to poorer families, and to have been preserved mainly there. But around the 1950s—presumably with the advent of roads and urbanization and the move away from living on remote farms—the tradition seems to have all but died out. (In rural schools in southern Brazil there were however apparently girls in the 1950s being taught in art classes how to make folk Spikeys with openings in them—to serve as piggy banks.)

Folk Spikeys seem to have shown up with different stories in different places around Brazil. In the southern border region (near Argentina and Uruguay) there's apparently a tradition that the "Star of St. Miguel" (aka folk Spikey) was made in villages by healer women (aka "witches"), who were supposed to think about the health of the person being healed while they were sewing their Spikeys.

In other parts of Brazil, folk Spikeys sometimes seem to be referred to by the names of flowers and fruits that look vaguely similar. In the northeast, "Flor Mandacarú" (after flowers on a cactus). In tropical wetland areas, "Carambola" (after star fruit). And in central forest areas "Pindaíva" (after a spiky red fruit).

But the most common current name for a folk Spikey seems to be "Giramundo"—an apparently not-very-recent Portuguese constructed word meaning essentially "whirling world." The folk Spikey, it seems, was used like a charm, and was supposed to bring good luck as it twirled in the wind. The addition of tails seems to be recent, but apparently it was common to hang up folk Spikeys in houses, perhaps particularly on festive occasions.

It's often not clear what's original, and what's a more recent tradition that happens to have "entrained" folk Spikeys. In the Three Kings' Day parade (as in the three kings from the Bible) in São Luiz do Paraitinga, folk Spikeys are apparently used to signify the Star of Bethlehem—but this seems to just be a recent thing, definitely not indicative of some ancient religious connection.

We've found a couple of examples of folk Spikeys showing up in art exhibitions. One was in a 1963 exhibition about folk art from northeastern Brazil organized by architect Lina Bo Bardi. The other, which happens to be the largest 3D Spikey I've ever seen, was in a 1997 exhibition of work by architect and set designer Flávio Império:

So... where did the folk Spikey come from? I still don't know. It may have originated in Brazil; it may have come from Portugal or elsewhere in Europe. The central use of fabrics and sewing needed to make a "60° Spikey" work might argue against an Amerindian or African origin.

One modern Spikey artisan did say that her great-grandmother—who made folk Spikeys and was born in the late 1800s—came from the Romagna region of Italy. (One also said she learned about folk Spikeys from her French-Canadian grandmother.) And I suppose it's conceivable that at one time there were folk Spikeys all over Europe, but they died out enough generations ago that no oral tradition about them survives. Still, while a decent number of polyhedra appear, for example, in European paintings from earlier centuries, I don't know of a single Spikey among them. (I also don't know of any Spikeys in historical Islamic art.)

But ultimately I'm pretty sure that somewhere there's a single origin for the folk Spikey. It's not something that I suspect was invented more than once.

I have to say that I've gone on "art origin hunts" before. One of the more successful was looking for the first nested (Sierpiński) pattern—which eventually led me to a crypt in a church in Italy, where I could see the pattern being progressively discovered, in signed stone mosaics from just after the year 1200.

So far the Spikey has proved more elusive—and it certainly doesn't help that the primary medium in which it appears to have been explored involved fabric, which doesn't keep the way stone does.

Spikeys Come to Life

Whatever its ultimate origins, Spikey serves us very well as a strong and dignified icon. But sometimes it's fun to have Spikey "come to life"—and over the years we've made various "personified Spikeys" for various purposes:

When you use Wolfram|Alpha, it'll usually show its normal, geometrical Spikey. But just sometimes your query will make the Spikey "come to life"—as it does for pi queries on Pi Day:

Spikeys Forever

Polyhedra are timeless. You see a polyhedron in a picture from 500 years ago and it'll look just as clean and modern as a polyhedron from my computer today.

I've spent a fair fraction of my life finding abstract, computational things (think cellular automaton patterns). And they too have a timelessness to them. But—try as I might—I have not found much of a thread of history for them. As abstract objects they could have been created at any time. But in fact they are modern, created because of the conceptual framework we now have, and with the tools we have today—and never seen before.

Polyhedra have both timelessness and a rich history that goes back thousands of years. In their appearance, polyhedra remind us of gems. And finding a certain kind of regular polyhedron is a bit like finding a gem out in the geometrical universe of all possible shapes.

The rhombic hexecontahedron is a wonderful such gem, and as I have explored its properties, I have come to have even more appreciation for it.

But it is also a gem with a human story—and it is so interesting to see how something as abstract as a polyhedron can connect people across the world with such diverse backgrounds and objectives.

Who first came up with the rhombic hexecontahedron? We don't know, and perhaps we never will. But now that it is here, it's forever. My favorite polyhedron.

Advance of the Data Civilization: A Timeline

August 16, 2011

The precursors of what we're trying to do with computable data in Wolfram|Alpha in many ways stretch back to the very dawn of human history—and in fact their development has been fascinatingly tied to the whole progress of civilization.

Last year we invited the leaders of today's great data repositories to our Wolfram Data Summit—and as a conversation piece we assembled a timeline of the historical development of systematic data and computable knowledge.

This year, as we approach the Wolfram Data Summit 2011, we've taken the comments and suggestions we got, and we're making available a five-feet-long (1.5 meters) printed poster of the timeline—as well as having the basic content on the web.

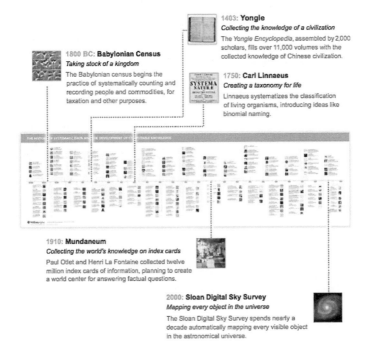

1403: Yongle
Collecting the knowledge of a civilization
The *Yongle Encyclopedia*, assembled by 2,000 scholars, fills over 11,000 volumes with the collected knowledge of Chinese civilization.

1800 BC: Babylonian Census
Taking stock of a kingdom
The Babylonian census begins the practice of systematically counting and recording people and commodities, for taxation and other purposes.

1750: Carl Linnaeus
Creating a taxonomy for life
Linnaeus systematizes the classification of living organisms, introducing ideas like binomial naming.

1910: Mundaneum
Collecting the world's knowledge on index cards
Paul Otlet and Henri La Fontaine collected twelve million index cards of information, planning to create a world center for answering factual questions.

2000: Sloan Digital Sky Survey
Mapping every object in the universe
The Sloan Digital Sky Survey spends nearly a decade automatically mapping every visible object in the astronomical universe.

The story the timeline tells is a fascinating one: of how, in a multitude of steps, our civilization has systematized more and more areas of knowledge—collected the data associated with them, and gradually made them amenable to automation.

The usual telling of history makes scant mention of most of these developments—though so many of them are so obvious in our lives today. Weights and measures. The calendar. Alphabetical lists. Plots of data. Dictionaries. Maps. Music notation. Stock charts. Timetables. Public records. ZIP Codes. Weather reports. All the things that help us describe and organize our world.

Historically, each one required an idea, and had an origin. Most often, what was happening was that some aspect of the world was effectively getting bigger—and one organization or one person took the lead in introducing a method of systematization.

Sometimes those involved were powerful or famous. But quite often they were in a sense in a back room, just solving a practical problem— usually modestly at first. Yet in time the perhaps arbitrary schemes they invented gradually spread as the need for them increased.

Most people will have heard of Euclid, who defined a way to systematize mathematics, or of Julius Caesar, who standardized the months of the year. Fewer will have heard of Guido d'Arezzo, who in 1030 AD invented stave notation for music. Or Robert Cawdrey, who in 1604 made what was probably the first alphabetical dictionary. Or Munehisa Homma, who in 1755 made what was probably the first market price chart. Or George Bradshaw, who in 1839 made the first train timetable. Or Malcolm Dyson, who in 1946 invented the standard IUPAC notation for naming chemicals.

As one looks at the whole timeline, one can see several definite classes of innovations.

One class are schemes for describing or representing things. Like latitude/longitude (invented by Eratosthenes around 200 BC). Or the notation for algebra (from Franciscus Vieta around 1595). Or binomial species names (invented by Carl Linnaeus around 1750). Or geological periods (introduced around 1830). Or citations for legal cases (from

Frank Shepard in 1873). Or CIE color space (from 1931). Or SI units (from 1954). Or ASCII code (from 1963). Or DNS for internet addresses (from 1983).

Another class of innovations are schemes or repositories for collecting knowledge about things. Like Babylonian land records (from 3000 BC). Or the Library at Thebes (from 1250 BC). Or Ptolemy's star catalog (from 150 AD). Or the *Yongle Encyclopedia* (from 1403). Or the US Census (from 1790). Or *Who's Who* (from 1849). Or weather charts (from Robert FitzRoy in 1860). Or the *Oxford English Dictionary* (from the 1880s). Or the "Yellow Pages" (from Reuben H. Donnelly in 1886). Or *Chemical Abstracts* (from 1907). Or baseball statistics (from Al Elias in 1913). Or Gallup polls (from 1935). Or GenBank (from 1982).

Another class of innovations are more abstract: in effect formalisms for handling knowledge. Like arithmetic (from 20,000 BC). Or formal grammar (from Panini around 400 BC). Or logic (from Aristotle around 350 BC). Or demographic statistics (notably from John Graunt in 1662). Or calculus (from Isaac Newton and Gottfried Leibniz around 1687). Or flow charts (from Frank & Lillian "Cheaper by the Dozen" Gilbreth in 1921). Or computer languages (from around 1957). Or geographic information systems (from Roger Tomlinson in 1962). Or relational databases (from the 1970s).

And then, of course, there is the curious history of attempts to do things like what Wolfram|Alpha does. I suppose Aristotle was already thinking of something similar around 350 BC, as he tried to classify objects in the world, and use logic to formalize reasoning. And then in the 1680s there was Gottfried Leibniz, who very explicitly wanted to convert all human questions to a universal symbolic language, and use a logic-based machine to get answers—with knowledge ultimately coming from libraries he hoped to assemble.

Needless to say, both Aristotle and Leibniz lived far too early to make these things work. But occasionally the ideas reemerged. And for example starting around 1910 Paul Otlet and Henri La Fontaine actually collected 12 million index cards of information for their Mundaneum, with the idea of operating a telegraph-based world question-answering center.

In 1937 H. G. Wells presented his vision for a "world brain", and in 1945 Vannevar Bush described his "memex", that would give computerized access to the world's knowledge. And by the 1950s and 1960s, it began to be taken almost for granted that knowledge would someday become computable—as portrayed in movies like *Desk Set* or *2001: A Space Odyssey*, or in television shows like *Star Trek*.

The assumption, however, was that the key innovation would be "artificial intelligence"—an automation of human intelligence. And as the years went by, and artificial intelligence languished, so too did progress in making knowledge broadly computable.

As I've talked about elsewhere, my own key realization—that arose from my basic research in *A New Kind of Science*—is that there can't ever ultimately be anything special about intelligence: it's all just computation. But where should the raw material for that computation come from? The point is that it does not have to be learned, as a human would, through some incremental process of education. Rather, we can just start from the whole corpus of systematic knowledge and data—as well as methods and models and algorithms—that our civilization has accumulated, poured wholesale into our computational system.

And this is what we have done with Wolfram|Alpha: in effect making immediate direct use of the whole rich history portrayed in the timeline.

I should say that as a person interested in the history of ideas, the actual process of assembling the timeline was a quite fascinating one. We started by looking at all the different areas of knowledge that we cover in Wolfram|Alpha—or hope to cover. Then in effect we worked backward, trying to find the earliest historical antecedents that defined each area.

Sometimes most of us knew these antecedents. But quite often we were surprised by how long ago—or how recent—those antecedents actually were. And in some cases we had to ask a whole string of experts before we were confident that we had the right story.

Each entry on the timeline was written separately—and I was most curious to see what would emerge when the whole timeline was put together. Of course, there is considerable arbitrariness to what actually appears on the timeline, and inevitably it's prejudiced toward more recent developments, not least because these do not have to have survived as long to seem important today.

But when I first looked at the completed timeline, the first thing that struck me was how much two entities stood out in their contributions: ancient Babylon, and the United States government. For Babylon—as the first great civilization—brought us such things as the first known census, standardized measures, the calendar, land registration, codes of laws, and the first known mathematical tables. In the United States, perhaps it was the spirit of building a country from scratch, or perhaps the notion of "government for the people", but starting as early as 1785 (with the formation of the US Land Ordinance), the US government began an impressive series of firsts in systematic data collection.

Given the timeline, a very obvious question is: how are all these events distributed in time, and space?

Here's a plot showing the number of events per decade and per century:

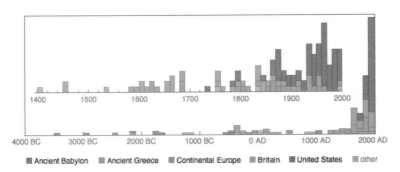

And here's a cumulative version of the same information:

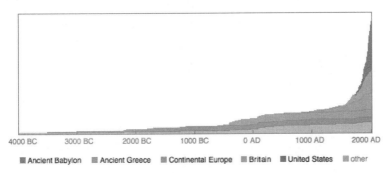

In the first plot, we see a burst of activity in the golden age of Ancient Greece. And then we see more in the Renaissance, the Industrial Revolution, and the Computer Revolution. But it is notable that there is still at least some activity even in Europe in the Middle Ages.

Looking at the cumulative plot, we see the center of activity shift from Babylon to Greece around 500 BC, then to continental Europe around 1000 AD (after modest activity in the Roman Empire). Around 1600 Britain begins to take off, firmly rivaling continental Europe by the mid-1800s. The US starts to show activity before 1800, but really takes off in the early 1900s.

Here's how the share of "events so far" evolves over time:

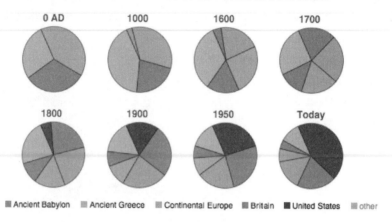

Ancient Greece surpasses Babylon in 250 BC. Europe surpasses Greece in 1595. Britain briefly surpasses continental Europe in 1786. The US surpasses Britain in 1942, and all of Europe in 1984—and today is only 12% short of surpassing everything before it put together.

It's notable how concentrated everything is in the typical "Western Civilization" countries. Perhaps this reflects our ignorance of other history, but I rather suspect it reflects instead the different interests of different cultures—and their different approaches to knowledge.

One of the most obvious features of the plots above is the rapid acceleration of entries in recent times. As I mentioned before, there's inevitably a survival bias. But to me what's somewhat remarkable is that nearly 20% of what's on the timeline was already done by 1000 AD, 40% by 1800 and

60% by 1900. If one looks at the last 500 years, though, there's a surprisingly good fit to an exponential increase, doubling every 95 years.

Now remember, the timeline is not about technology or science, it's about data and knowledge. When you look at the timeline, you might ask: "Where's Einstein? Where's Darwin? Where's the space program?" Well, they're not there. Because despite their importance in the history of science and technology, they're not really part of the particular story the timeline is telling: of how systematic data and knowledge came to be the way it is in our world. And as I said before, much of this is "back room history", not really told in today's history books.

In Wolfram|Alpha, we also have a growing amount of information about more traditional science/technology inventions and discoveries. And the timeline for these looks a little different. There is much less activity in the Middle Ages, for example, and in the last 500 years, there is growth that rather noisily fits as exponential, with a 75-year doubling time. If anything, there are even more dramatic survival bias effects here than in the data+knowledge timeline. But if there is a significance to the difference between the timelines, perhaps it reflects the fact that the systematization of data and knowledge provides core infrastructure for the world—and grows more slowly and steadily, gradually making possible all those other innovations.

In any case, as we work on Wolfram|Alpha, it is sobering to see how long the road to where we are today has been. But it is exciting to see how much further modern technology has already made it possible for us to go. And I am proud to be a small part of such a distinguished and long history.

Data Science of the Facebook World

April 24, 2013

More than a million people have now used our Wolfram|Alpha Personal Analytics for Facebook. And as part of our latest update, in addition to collecting some anonymized statistics, we launched a Data Donor program that allows people to contribute detailed data to us for research purposes.

A few weeks ago we decided to start analyzing all this data. And I have to say that if nothing else it's been a terrific example of the power of Mathematica and the Wolfram Language for doing data science. (It'll also be good fodder for the data science course I'm starting to create.)

We'd always planned to use the data we collect to enhance our Personal Analytics system. But I couldn't resist also trying to do some basic science with it.

I've always been interested in people and the trajectories of their lives. But I've never been able to combine that with my interest in science. Until now. And it's been quite a thrill over the past few weeks to see the results we've been able to get. Sometimes confirming impressions I've had, sometimes showing things I never would have guessed, and all along reminding me of phenomena I've studied scientifically in *A New Kind of Science*.

So what does the data look like? Here are the social networks of a few Data Donors—with clusters of friends given different colors. (Anyone can find their own network using Wolfram|Alpha—or the SocialMediaData function in Mathematica.)

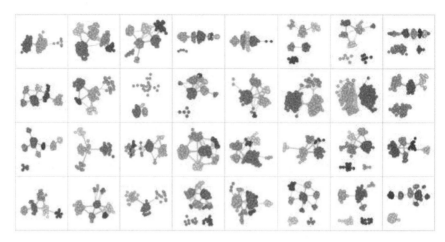

So a first quantitative question to ask is: How big are these networks usually? In other words, how many friends do people typically have on Facebook? Well, at least for our users, that's easy to answer. The median is 342—and here's a histogram showing the distribution (there's a cutoff at 5000 because that's the maximum number of friends for a personal Facebook page):

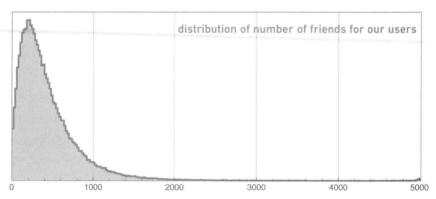

But how typical are our users? In most respects—so far as we can tell—they seem pretty typical. But there are definitely some differences. Like here's the distribution of the number of friends not just for our users, but also for their friends (there's a mathematical subtlety in deriving this that I'll discuss later):

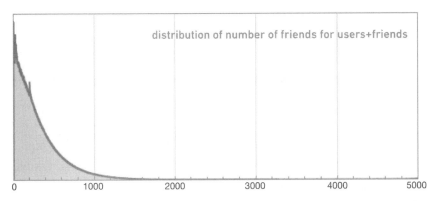

And what we see is that in this broader Facebook population, there are significantly more people who have almost no Facebook friends. Whether such people should be included in samples one takes is a matter of debate. But so long as one looks at appropriate comparisons, aggregates, and so on, they don't seem to have a huge effect. (The spike at 200 friends probably has to do with Facebook's friend recommendation system.)

So, OK. Let's ask for examples of how the typical number of Facebook friends varies with a person's age. Of course all we know are self-reported "Facebook ages." But let's plot how the number of friends varies with that age. The solid line is the median number of friends; successive bands show successive octiles of the distribution.

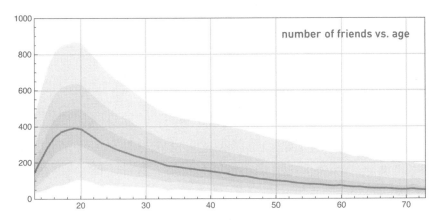

After a rapid rise, the number of friends peaks for people in their late teenage years, and then declines thereafter. Why is this? I suspect it's partly a reflection of people's intrinsic behavior, and partly a reflection of the fact that Facebook hasn't yet been around very long. Assuming people don't drop friends much once they've added them one might expect that the number of friends would simply grow with age. And for sufficiently young people that's basically what we see. But there's a limit to the growth, because there's a limit to the number of years people have been on Facebook. And assuming that's roughly constant across ages, what the plot suggests is that people add friends progressively more slowly with age.

But what friends do they add? Given a person of a particular age, we can for example ask what the distribution of ages of the person's friends is. Here are some results (the jaggedness, particularly at age 70, comes from the limited data we have):

friend ages for people of different ages

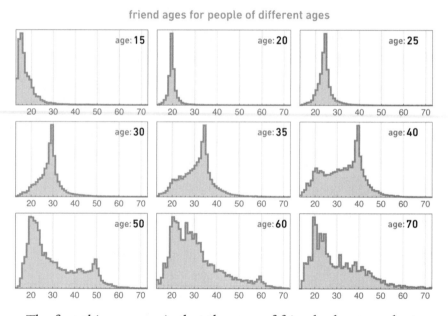

The first thing we see is that the ages of friends always peak at or near the age of the person themselves—which is presumably a reflection of the fact that in today's society many friends are made in age-based classes in school or college. For younger people, the peak around the person's age tends to be pretty sharp. For older people, the distribution gets progressively broader.

We can summarize what happens by plotting the distribution of friend ages against the age of a person (the solid line is the median age of friends):

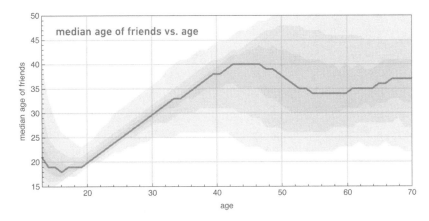

There's an anomaly for the youngest ages, presumably because of kids under 13 misreporting their ages. But apart from that, we see that young people tend to have friends who are remarkably close in age to themselves. The broadening as people get older is probably associated with people making non-age-related friends in their workplaces and communities. And as the array of plots above suggests, by people's mid-40s, there start to be secondary peaks at younger ages, presumably as people's children become teenagers, and start using Facebook.

So what else can one see about the trajectory of people's lives? Here's the breakdown according to reported relationship status as a function of age:

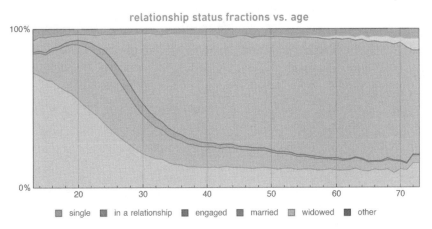

And here's more detail, separating out fractions for males and females ("married+" means "civil union," "separated," "widowed," etc. as well as "married"):

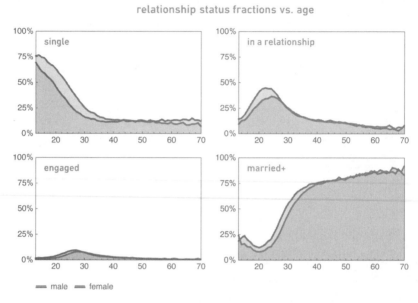

relationship status fractions vs. age

There's some obvious goofiness at low ages with kids (slightly more often girls than boys) misreporting themselves as married. But in general the trend is clear. The rate of getting married starts going up in the early 20s—a couple of years earlier for women than for men—and decreases again in the late 30s, with about 70% of people by then being married. The fraction of people "in a relationship" peaks around age 24, and there's a small "engaged" peak around 27. The fraction of people who report themselves as married continues to increase roughly linearly with age, gaining about 5% between age 40 and age 60—while the fraction of people who report themselves as single continues to increase for women, while decreasing for men.

I have to say that as I look at the plots above, I'm struck by their similarity to plots for physical processes like chemical reactions. It's as if all those humans, with all the complexities of their lives, still behave in aggregate a bit like molecules—with certain "reaction rates" to enter into relationships, marry, etc.

Of course, what we're seeing here is just for the "Facebook world." So how does it compare to the world at large? Well, at least some of what we can measure in the Facebook world is also measured in official censuses. And so for example we can see how our results for the fraction of people married at a given age compare with results from the official US Census:

fraction married vs. age

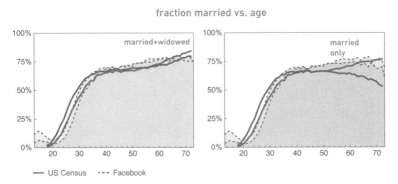

I'm amazed at how close the correspondence is. Though there are clearly some differences, like below-age-20 kids on Facebook misreporting themselves as married. And on the older end, widows are still considering themselves married for purposes of Facebook. For people in their 20s, there's also a small systematic difference—with people on Facebook on average getting married a couple of years later than the Census would suggest. (As one might expect, if one excludes the rural US population, the difference gets significantly smaller.)

Talking of the Census, we can ask in general how our Facebook population compares to the US population. And for example, we find, not surprisingly, that our Facebook population is heavily weighted toward younger people:

population vs. age

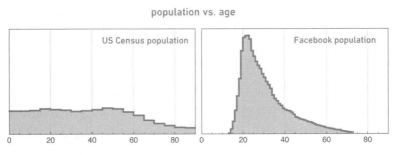

OK. So we saw above how the typical number of friends a person has depends on age. What about gender? Perhaps surprisingly, if we look at all males and all females, there isn't a perceptible difference in the distributions of number of friends. But if we instead look at males and females as a function of age, there is a definite difference:

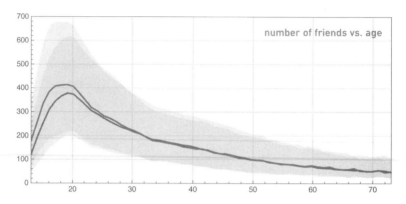

Teenage boys tend to have more friends than teenage girls, perhaps because they are less selective in who they accept as friends. But after the early 20s, the difference between genders rapidly dwindles.

What effect does relationship status have? Here's the male and female data as a function of age:

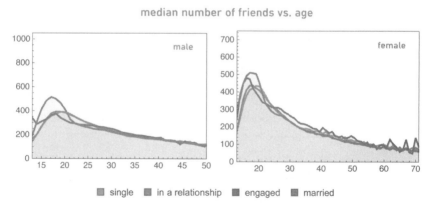

In the older set, relationship status doesn't seem to make much difference. But for young people it does, with teenagers who (mis)report themselves as "married" on average having more friends than those who don't. And with early teenage girls who say they're "engaged" (perhaps to be able to tag a BFF) typically having more friends than those who say they're single, or just "in a relationship."

Another thing that's fairly reliably reported by Facebook users is location. And it's common to see quite a lot of variation by location. Like here are comparisons of the median number of friends for countries around the world (ones without enough data are the lightest gray), and for states in the US:

median number of friends by location

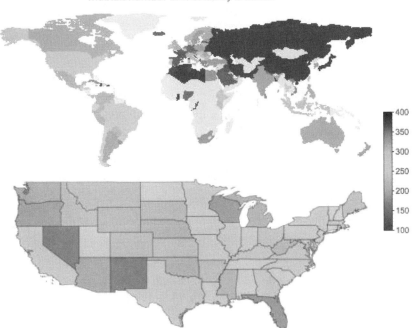

There are some curious effects. Countries like Russia and China have low median friend counts because Facebook isn't widely used for connections between people inside those countries. And perhaps there are lower friend counts in the western US because of lower population densities. But quite why there are higher friend counts for our Facebook population in places like Iceland, Brazil, and the Philippines—or Mississippi—I don't know. (There is of course some "noise" from people misreporting their locations. But with the size of the sample we have, I don't think this is a big effect.)

In Facebook, people can list both a "hometown" and a "current city." Here's how the probability that these are in the same US state varies with age:

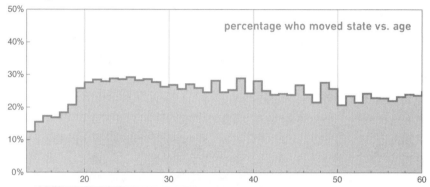

What we see is pretty much what one would expect. For some fraction of the population, there's a certain rate of random moving, visible here for young ages. Around age 18, there's a jump as people move away from their "hometowns" to go to college and so on. Later, some fraction move back, and progressively consider wherever they live to be their "hometown."

One can ask where people move to and from. Here's a plot showing the number of people in our Facebook population moving between different US states and different countries:

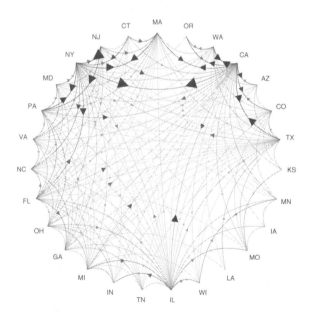

There's a huge range of demographic questions we could ask. But let's come back to social networks. It's a common observation that people tend to be friends with people who are like them. So to test this we might for example ask whether people with more friends tend to have friends who have more friends. Here's a plot of the median number of friends that our users have, as a function of the number of friends that they themselves have:

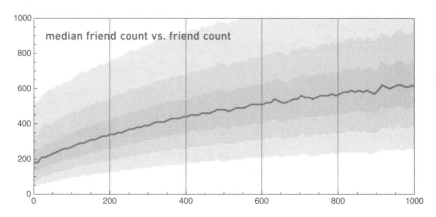

And the result is that, yes, on average people with more friends tend to have friends with more friends. Though we also notice that people with lots of friends tend to have friends with fewer friends than themselves.

And seeing this gives me an opportunity to discuss a subtlety I alluded to earlier. The very first plot in this chapter shows the distribution of the number of friends that our users have. But what about the number of friends that their friends have? If we just average over all the friends of all our users, this is how what we get compares to the original distribution for our users themselves:

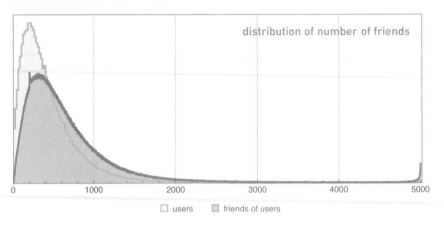

distribution of number of friends

☐ users   ▦ friends of users

It seems like our users' friends always tend to have more friends than our users themselves. But actually from the previous plot we know this isn't true. So what's going on? It's a slightly subtle but general social network phenomenon known as the "friendship paradox." The issue is that when we sample the friends of our users, we're inevitably sampling the space of all Facebook users in a very non-uniform way. In particular, if our users represent a uniform sample, any given friend will be sampled at a rate proportional to how many friends they have—with the result that people with more friends are sampled more often, so the average friend count goes up.

It's perfectly possible to correct for this effect by weighting friends in inverse proportion to the number of friends they have—and that's what we did earlier in this chapter. And by doing this we determine that in fact the friends of our users do not typically have more friends than our users themselves; instead their median number of friends is actually 229 instead of 342.

It's worth mentioning that if we look at the distribution of number of friends that we deduce for the Facebook population, it's a pretty good fit to a power law, with exponent −2.8. And this is a common form for networks of many kinds—which can be understood as the result of an effect known as "preferential attachment," in which as the network grows, nodes that already have many connections preferentially get more connections, leading to a limiting "scale-free network" with power-law features.

But, OK. Let's look in more detail at the social network of an individual user. I'm not sufficiently diligent on Facebook for my own network to be interesting. But my 15-year-old daughter Catherine was kind enough to let me show her network:

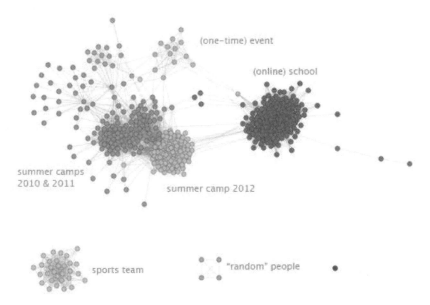

There's a dot for each of Catherine's Facebook friends, with connections between them showing who's friends with whom. (There's no dot for Catherine herself, because she'd just be connected to every other dot.) The network is laid out to show clusters or "communities" of friends (using the Wolfram Language function FindGraphCommunities). And it's amazing the extent to which the network "tells a story," with each cluster corresponding to some piece of Catherine's life or history.

Here's a whole collection of networks from our Data Donors:

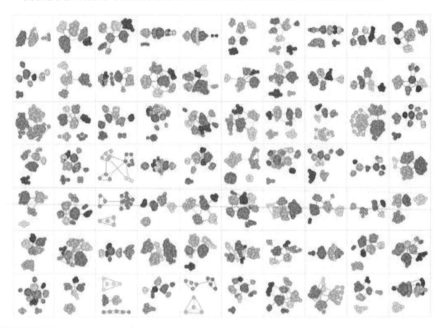

No doubt each of these networks tells a different story. But we can still generate overall statistics. Like, for example, here is a plot of how the number of clusters of friends varies with age (there'd be less noise if we had more data):

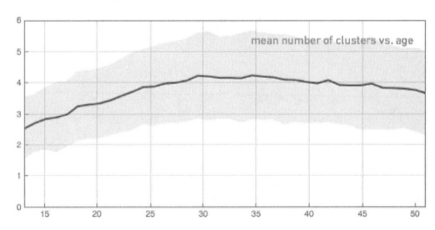

Even at age 13, people typically seem to have about three clusters (perhaps school, family, and neighborhood). As they get older, go to different schools, take jobs, and so on, they accumulate another cluster or so. Right now the number saturates above about age 30, probably in large part just because of the limited time Facebook has been around.

How big are typical clusters? The largest one is usually around 100 friends; the plot below shows the variation of this size with age:

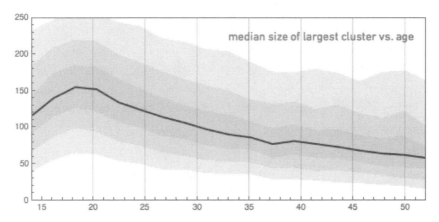

And here's how the size of the largest cluster as a fraction of the whole network varies with age:

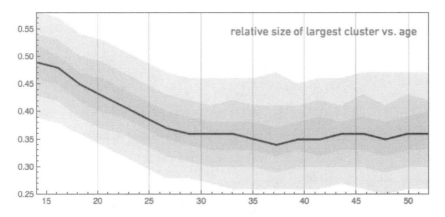

What about more detailed properties of networks? Is there a kind of "periodic table" of network structures? Or a classification scheme like the one I made long ago for cellular automata?

The first step is to find some kind of iconic summary of each network, which we can do for example by looking at the overall connectivity of clusters, ignoring their substructure. And so, for example, for Catherine (who happened to suggest this idea), this reduces her network to the following "cluster diagram":

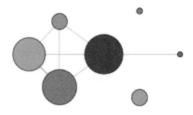

Doing the same thing for the Data Donor networks shown above, here's what we get:

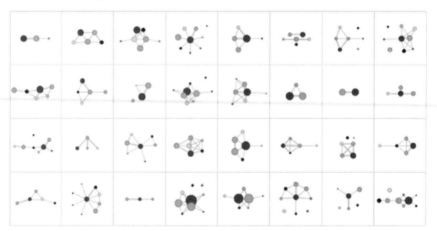

In making these diagrams, we're keeping every cluster with at least two friends. But to get a better overall view, we can just drop any cluster with, say, less than 10% of all friends—in which case for example Catherine's cluster diagram becomes just:

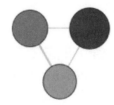

And now for example we can count the relative numbers of different types of structures that appear in all the Data Donor networks:

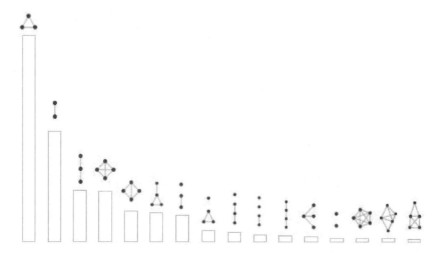

And we can look at how the fractions of each of these structures vary with age:

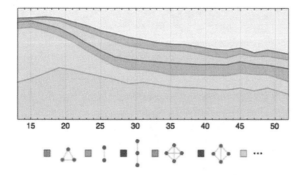

What do we learn? The most common structures consist of either two or three major clusters, all of them connected. But there are also structures in which major clusters are completely disconnected—presumably reflecting facets of a person's life that for reasons of geography or content are also completely disconnected.

For everyone there'll be a different detailed story behind the structure of their cluster diagram. And one might think this would mean that there could never be a general theory of such things. At some level it's a bit like trying to find a general theory of human history, or a general theory of the progression of biological evolution. But what's interesting now about the Facebook world is that it gives us so much more data from which to form theories.

And we don't just have to look at things like cluster diagrams, or even friend networks: we can dig almost arbitrarily deep. For example, we can analyze the aggregated text of posts people make on their Facebook walls, say classifying them by topics they talk about (this uses a natural language classifier written in the Wolfram Language and trained using some large corpora):

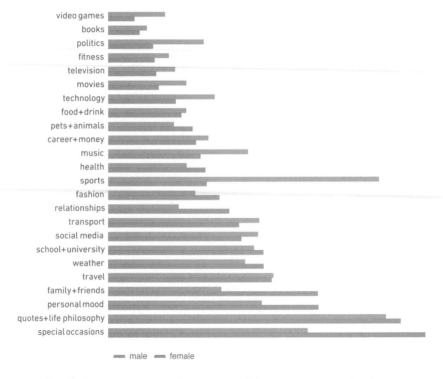

Each of these topics is characterized by certain words that appear with high frequency:

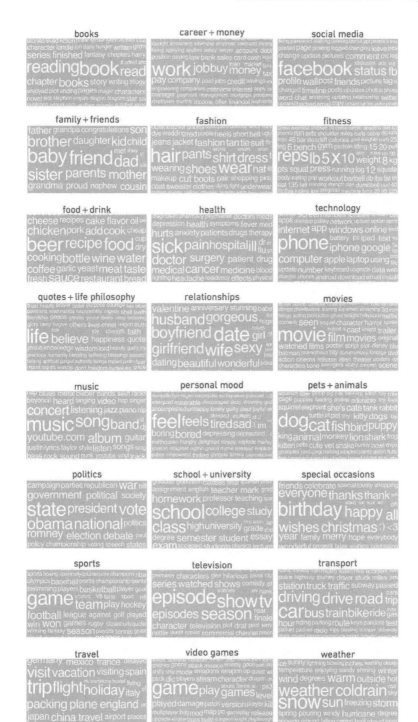

And for each topic we can analyze how its popularity varies with (Facebook) age:

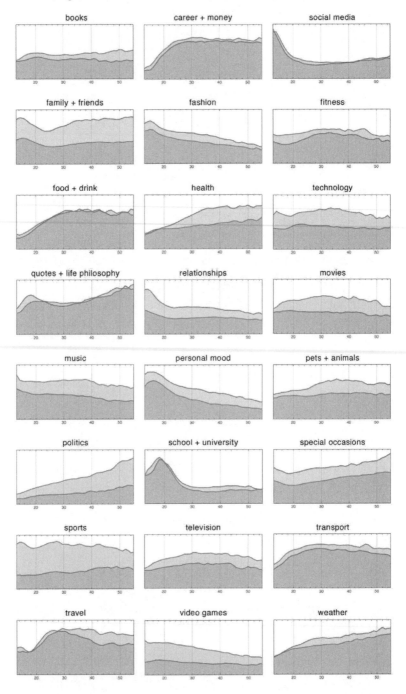

It's almost shocking how much this tells us about the evolution of people's typical interests. People talk less about video games as they get older, and more about politics and the weather. Men typically talk more about sports and technology than women—and, somewhat surprisingly to me, they also talk more about movies, television, and music. Women talk more about pets+animals, family+friends, relationships—and, at least after they reach childbearing years, health. The peak time for anyone to talk about school+university is (not surprisingly) around age 20. People get less interested in talking about "special occasions" (mostly birthdays) through their teens, but gradually gain interest later. And people get progressively more interested in talking about career+money in their 20s. And so on, and so on.

Some of this is rather depressingly stereotypical. And most of it isn't terribly surprising to anyone who's known a reasonable diversity of people of different ages. But what to me is remarkable is how we can see everything laid out in such quantitative detail in the pictures above—kind of a signature of people's thinking as they go through life.

Of course, these pictures are all based on aggregate data, carefully anonymized. But if we start looking at individuals, we'll see all sorts of other interesting things. And for example personally I'm very curious to analyze my own archive of nearly 25 years of email—and then perhaps predict things about myself by comparing to what happens in the general population.

Over the decades I've been steadily accumulating countless anecdotal "case studies" about the trajectories of people's lives—from which I've certainly noticed lots of general patterns. But what's amazed me about what we've done over the past few weeks is how much systematic information it's been possible to get all at once. Quite what it all means, and what kind of general theories we can construct from it, I don't yet know.

But it feels like we're starting to be able to train a serious "computational telescope" on the "social universe." And it's letting us discover all sorts of phenomena that have the potential to help us understand much more about society and about ourselves. And that, by the way, provide great examples of what can be achieved with data science, and with the technology I've been working on developing for so long.

A Short Talk on AI Ethics

October 17, 2016

My mother was a philosophy professor in Oxford. And when I was a kid I always said the one thing I'd never do was do or talk about philosophy. But, well, here I am.

Before I really get into AI, I think I should say a little bit about my worldview. I've basically spent my life alternating between doing basic science and building technology. I've been interested in AI for about as long as I can remember. But as a kid I started out doing physics and cosmology and things. That got me into building technology to automate stuff like math. And that worked so well that I started thinking about how to really know and compute everything about everything. That was in about 1980—and at first I thought I had to build something like a brain, and I was studying neural nets and so on. But I didn't get too far.

And meanwhile I got interested in an even bigger problem in science: how to make the most general possible theories of things. The dominant idea for 300 years had been to use math and equations. But I wanted to go beyond them. And the big thing I realized was that the way to do that was to think about programs, and the whole computational universe of possible programs.

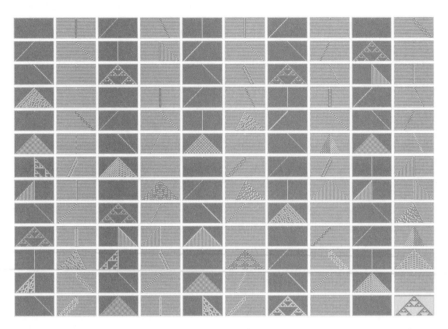

And that led to my personal Galileo-like moment. I just pointed my "computational telescope" at these simplest possible programs, and I saw this amazing one I called rule 30—that just seemed to go on producing complexity forever from essentially nothing.

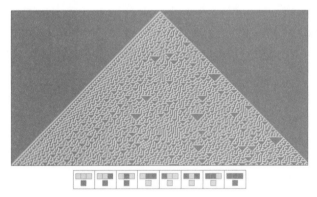

Well, after I'd seen this, I realized this is actually something that happens all over the computational universe—and all over nature. It's really the secret that lets nature make all the complicated stuff we see. But it's something else too: it's a window into what raw, unfettered computation is like. At least traditionally when we do engineering we're always building things that are simple enough that we can foresee what they'll do.

But if we just go out into the computational universe, things can be much wilder. Our company has done a lot of mining out there, finding programs that are useful for different purposes, like rule 30 is for randomness. And modern machine learning is kind of part way from traditional engineering to this kind of free-range mining.

But, OK, what can one say in general about the computational universe? Well, all these programs can be thought of as doing computations. And years ago I came up with what I call the Principle of Computational Equivalence—that says that if behavior isn't obviously simple, it typically corresponds to a computation that's maximally sophisticated. There are lots of predictions and implications of this. Like that universal computation should be ubiquitous. As should undecidability. And as should what I call computational irreducibility.

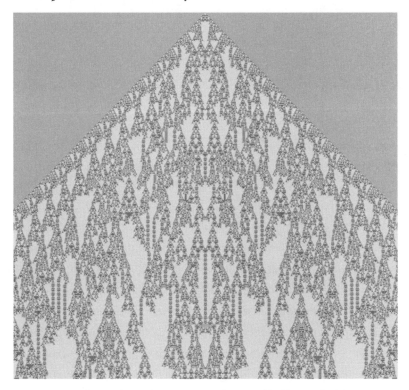

Can you predict what it's going to do? Well, it's probably computationally irreducible, which means you can't figure out what it's going to do without effectively tracing every step and going through the same computational effort it does. It's completely deterministic. But to us it's got what seems like free will—because we can never know what it's going to do.

Here's another thing: what's intelligence? Well, our big unifying principle says that everything—from a tiny program, to our brains, is computationally equivalent. There's no bright line between intelligence and mere computation. The weather really does have a mind of its own: it's doing computations just as sophisticated as our brains. To us, though, it's pretty alien computation. Because it's not connected to our human goals and experiences. It's just raw computation that happens to be going on.

So how do we tame computation? We have to mold it to our goals. And the first step there is to describe our goals. And for the past 30 years what I've basically been doing is creating a way to do that.

I've been building a language—that's now called the Wolfram Language—that allows us to express what we want to do. It's a computer language. But it's not really like other computer languages. Because instead of telling a computer what to do in its terms, it builds in as much knowledge as possible about computation and the world, so that we humans can describe in our terms what we want, and then it's up to the language to get it done as automatically as possible.

This basic idea has worked really well, and in the form of Mathematica it's been used to make endless inventions and discoveries over the years. It's also what's inside Wolfram|Alpha, where the idea is to take pure natural language questions, understand them, and use the kind of curated knowledge and algorithms of our civilization to answer them. And, yes, it's a very classic AIish thing. And of course it's computed answers to billions and billions of questions from humans, for example inside Siri.

I had an interesting experience recently, figuring out how to use what we've built to teach computational thinking to kids. I was writing exercises for a book. At the beginning, it was easy: "make a program to do X." But later on, it was like "I know what to say in the Wolfram Language, but it's really hard to express in English." And of course that's why I just spent 30 years building the Wolfram Language.

English has maybe 25,000 common words; the Wolfram Language has about 5000 carefully designed built-in constructs—including all the latest machine learning—together with millions of things based on curated data. And the idea is that once one can think about something in the world computationally, it should be as easy as possible to express it in the Wolfram Language. And the cool thing is, it really works. Humans, including kids, can read and write the language. And so can computers. It's a kind of high-level bridge between human thinking, in its cultural context, and computation.

OK, so what about AI? Technology has always been about finding things that exist, and then taming them to automate the achievement of particular human goals. And in AI the things we're taming exist in the computational universe. Now, there's a lot of raw computation seething around out there—just as there's a lot going on in nature. But what we're interested in is computation that somehow relates to human goals.

So what about ethics? Well, maybe we want to constrain the computation, the AI, to only do things we consider ethical. But somehow we have to find a way to describe what we mean by that.

Well, in the human world, one way we do this is with laws. But so how do we connect laws to computations? We may call them "legal codes", but today laws and contracts are basically written in natural language. There've been simple computable contracts in areas like financial derivatives. And now one's talking about smart contracts around cryptocurrencies.

But what about the vast mass of law? Well, Leibniz—who died 300 years ago next month—was always talking about making a universal language to, as we would say now, express it all in a computable way. He was a few centuries too early, but I think now we're finally in a position to do this.

I wrote about this recently,* but let me try to summarize. With the Wolfram Language we've managed to express a lot of kinds of things in the world—like the ones people ask Siri about. And I think we're

---

* *www.wolfr.am/computational-law*

now within sight of what Leibniz wanted: to have a general symbolic discourse language that represents everything involved in human affairs.

I see it basically as a language design problem. Yes, we can use natural language to get clues, but ultimately we have to build our own symbolic language. It's actually the same kind of thing I've done for decades in the Wolfram Language. Take even a word like "plus." Well, in the Wolfram Language there's a function called Plus, but it doesn't mean the same thing as the word. It's a very specific version, that has to do with adding things mathematically. And as we design a symbolic discourse language, it's the same thing. The word "eat" in English can mean lots of things. But we need a concept—that we'll probably refer to as "eat"—that's a specific version, that we can compute with.

So let's say we've got a contract written in natural language. One way to get a symbolic version is to use natural language understanding— just like we do for billions of Wolfram|Alpha inputs, asking humans about ambiguities. Another way might be to get machine learning to describe a picture. But the best way is just to write in symbolic form in the first place, and actually I'm guessing that's what lawyers will be doing before too long.

And of course once you have a contract in symbolic form, you can start to compute about it, automatically seeing if it's satisfied, simulating different outcomes, automatically aggregating it in bundles, and so on. Ultimately the contract has to get input from the real world. Maybe that input is "born digital," like data about accessing a computer system, or transferring bitcoin. Often it'll come from sensors and measurements—and it'll take machine learning to turn into something symbolic.

Well, if we can express laws in computable form maybe we can start telling AIs how we want them to act. Of course it might be better if we could boil everything down to simple principles, like Asimov's Laws of Robotics, or utilitarianism or something.

But I don't think anything like that is going to work. What we're ultimately trying to do is to find perfect constraints on computation, but computation is something that's in some sense infinitely wild. The issue already shows up in Gödel's theorem. Like let's say we're looking at integers and we're trying to set up axioms to constrain them to just work the way we think they do. Well, what Gödel showed is that no finite set of axioms can ever achieve this. With any set of axioms you choose, there won't just be the ordinary integers; there'll also be other wild things.

And the phenomenon of computational irreducibility implies a much more general version of this. Basically, given any set of laws or constraints, there'll always be "unintended consequences." This isn't particularly surprising if one looks at the evolution of human law. But the point is that there's theoretically no way around it. It's ubiquitous in the computational universe.

Now I think it's pretty clear that AI is going to get more and more important in the world—and is going to eventually control much of the infrastructure of human affairs, a bit like governments do now. And like with governments, perhaps the thing to do is to create an AI Constitution that defines what AIs should do.

What should the Constitution be like? Well, it's got to be based on a model of the world, and inevitably an imperfect one, and then it's got to say what to do in lots of different circumstances. And ultimately what it's got to do is provide a way of constraining the computations that happen to be ones that align with our goals. But what should those goals be? I don't think there's any ultimate right answer. In fact, one can enumerate goals just like one can enumerate programs out in the computational universe. And there's no abstract way to choose between them.

But for us there's a way to choose. Because we have particular biology, and we have a particular history of our culture and civilization. It's taken us a lot of irreducible computation to get here. But now we're just at some point in the computational universe, that corresponds to the goals that we have.

Human goals have clearly evolved through the course of history. And I suspect they're about to evolve a lot more. I think it's pretty inevitable that our consciousness will increasingly merge with technology. And eventually maybe our whole civilization will end up as something like a box of a trillion uploaded human souls.

But then the big question is: "what will they choose to do?" Well, maybe we don't even have the language yet to describe the answer. If we look back even to Leibniz's time, we can see all sorts of modern concepts that hadn't formed yet. And when we look inside a modern machine learning or theorem proving system, it's humbling to see how many concepts it effectively forms—that we haven't yet absorbed in our culture.

Maybe looked at from our current point of view, it'll just seem like those disembodied virtual souls are playing videogames for the rest of eternity. At first maybe they'll operate in a simulation of our actual universe. Then maybe they'll start exploring the computational universe of all possible universes.

But at some level all they'll be doing is computation—and the Principle of Computational Equivalence says it's computation that's fundamentally equivalent to all other computation. It's a bit of a letdown. Our proud future ending up being computationally equivalent just to plain physics, or to little rule 30.

Of course, that's just an extension of the long story of science showing us that we're not fundamentally special. We can't look for ultimate meaning in where we've reached. We can't define an ultimate purpose. Or ultimate ethics. And in a sense we have to embrace the details of our existence and our history.

There won't be a simple principle that encapsulates what we want in our AI Constitution. There'll be lots of details that reflect the details of our existence and history. And the first step is just to understand how to represent those things. Which is what I think we can do with a symbolic discourse language.

And, yes, conveniently I happen to have just spent 30 years building the framework to create such a thing. And I'm keen to understand how we can really use it to create an AI Constitution.

Overcoming Artificial Stupidity

April 17, 2012

Today marks an important milestone for Wolfram|Alpha, and for computational knowledge in general: for the first time, Wolfram|Alpha is now on average giving complete, successful responses to more than 90% of the queries entered on its website (and with "nearby" interpretations included, the fraction is closer to 95%).

I consider this an impressive achievement—the hard-won result of many years of progressively filling out the knowledge and linguistic capabilities of the system.

The picture below shows how the fraction of successful queries (dark gray) has increased relative to unsuccessful ones (in light gray) since Wolfram|Alpha was launched in 2009. And from the log scale in the right-hand panel, we can see that there's been a roughly exponential decrease in the failure rate, with a half-life of around 18 months. It seems to be a kind of Moore's law for computational knowledge: the net effect of innumerable individual engineering achievements and new ideas is to give exponential improvement.

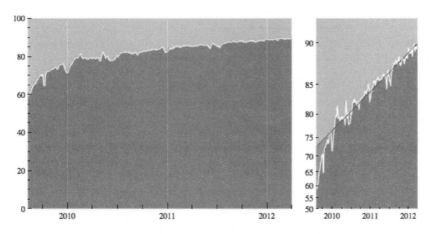

But to celebrate reaching our 90% query success rate, I thought it'd be fun to take a look at some of what we've left behind. Ever since the early days of Wolfram|Alpha, we've been keeping a scrapbook of our favorite examples of "artificial stupidity": places where Wolfram|Alpha gets the wrong idea, and applies its version of "artificial intelligence" to go off in what seems to us humans as a stupid direction.

Here's an example, captured over a year ago (and now long-since fixed):

When we typed "guinea pigs", we probably meant those furry little animals (which for example I once had as a kid). But Wolfram|Alpha somehow got the wrong idea, and thought we were asking about pigs in the country of Guinea, and diligently (if absurdly, in this case) told us that there were 86,431 of those in a 2008 count.

At some level, this wasn't such a big bug. After all, at the top of the output Wolfram|Alpha perfectly well told us it was assuming "'guinea' is a country", and offered the alternative of taking the input as a "species specification" instead. And indeed, if one tries the query today, the species is the default, and everything is fine. But having the wrong default interpretation a year ago was a simple but quintessential example of artificial stupidity, in which a subtle imperfection can lead to what seems to us laughably stupid behavior.

Here's what "guinea pigs" does today—a good and sensible result:

Here are some other examples from our scrapbook of artificial stupidity, collected over the past three years [2009–2011]. I'm happy to say that every single one of these now works nicely; many actually give rather impressive results.

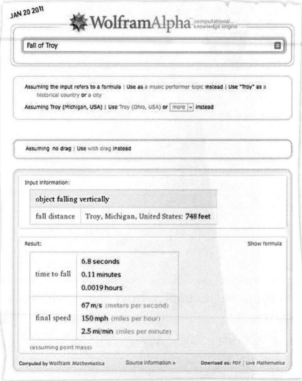

There's a certain humorous absurdity to many of these examples. In fact, looking at them suggests that this kind of artificial stupidity might actually be a good systematic source of things that we humans find humorous.

But where is the artificial stupidity coming from? And how can we overcome it?

There are two main issues that seem to combine to produce most of the artificial stupidity we see in these scrapbook examples. The first is that Wolfram|Alpha tries too hard to please—valiantly giving a result even if it doesn't really know what it's talking about. And the second is that it may simply not know enough—so that it misses the point because it's completely unaware of some possible meaning for a query.

Curiously enough, these two issues come up all the time for humans too—especially, say, when they're talking on a bad cellphone connection, and can't quite hear clearly.

For humans, we don't yet know the internal story of how these things work. But in Wolfram|Alpha it's very well defined. It's millions of lines of Mathematica code, but ultimately what Wolfram|Alpha does is to take the fragment of natural language it's given as input, and try to map it into some precise symbolic form (in the Mathematica language) that represents in a standard way the meaning of the input—and from which Wolfram|Alpha can compute results.

By now—particularly with data from nearly three years of actual usage—Wolfram|Alpha knows an immense amount about the detailed structure and foibles of natural language. And of necessity, it has to go far beyond what's in any grammar book.

When people type input into Wolfram|Alpha, I think we're seeing a kind of linguistic representation of undigested thoughts. It's not a random soup of words (as people might feed a search engine). It has structure—often quite complex—but it has scant respect for the niceties of traditional word order or grammar.

And as far as I am concerned one of the great achievements of Wolfram|Alpha is the creation of a linguistic understanding system that's robust enough to handle such things, and to successfully convert them to precise computable symbolic expressions.

One can think of any particular symbolic expression as having a certain "basin of attraction" of linguistic forms that will lead to it. Some of these forms may look perfectly reasonable. Others may look odd—but that doesn't mean they can't occur in the "stream of consciousness" of actual Wolfram|Alpha queries made by humans.

And usually it won't hurt anything to allow even very odd forms, with quite bizarre distortions of common language. Because the worst that will happen is that these forms just won't ever actually get used as input.

But here's the problem: what if one of those forms overlaps with something with a quite different meaning? If it's something that Wolfram|Alpha knows about, its linguistic understanding system will recognize the clash, and—if all is working properly—will choose the correct meaning.

But what happens if the overlap is with something Wolfram|Alpha doesn't know about?

In the last scrapbook example above Wolfram|Alpha was asked "what is a plum". At the time, it didn't know about fruits that weren't explicitly plant types. But it did happen to know about a crater on the Moon named "Plum". The linguistic understanding system certainly noticed the indefinite article "a" in front of "plum". But knowing nothing with the name "plum" other than a Moon crater (and erring—at least on the website—in the direction of giving some response rather than none), it concluded that the "a" must be some kind of "linguistic noise", went for the Moon crater meaning, and did something that looks to us quite stupid.

How can Wolfram|Alpha avoid this? The answer is simple: it just has to know more.

One might have thought that doing better at understanding natural language would be about covering a broader range of more grammar-like forms. And certainly this is part of it. But our experience with Wolfram|Alpha is that it is at least as important to add to the knowledgebase of the system.

A lot of artificial stupidity is about failing to have "common sense" about what an input might mean. Within some narrow domain of knowledge an interpretation might seem quite reasonable. But in a more general "common sense" context, the interpretation is obviously absurd. And the point is that as the domains of Wolfram|Alpha knowledge expand, they gradually fill out all the areas that we humans consider common sense, pushing out absurd "artificially stupid" interpretations.

Sometimes Wolfram|Alpha can in a sense overshoot. Consider the query "clever population". What does it mean? The linguistic construction seems a bit odd, but I'd probably think it was talking about how many clever people there are somewhere. But here's what Wolfram|Alpha says:

And the point is that Wolfram|Alpha knows something I don't: that there's a small city in Missouri named "Clever". Aha! Now the construction "clever population" makes sense. To people in southwestern Missouri, it would probably always have been obvious. But with typical everyday knowledge and common sense, it's not. And just like Wolfram|Alpha in the scrapbook examples above, most humans will assume that the query is about something completely different.

There've been a number of attempts to create natural language question-answering systems in the history of work on artificial intelligence. And in terms of immediate user impression, the problem with these systems has usually been not so much a failure to create artificial intelligence but rather the presence of painfully obvious artificial stupidity. In ways much more dramatic than these scrapbook examples, the system will "grab" a meaning it happens to know about, and robotically insist on using this, even though to a human it will seem stupid.

And what we learn from the Wolfram|Alpha experience is that the problem hasn't been our failure to discover some particular magic human-thinking-like language understanding algorithm. Rather, it's in a sense broader and more fundamental: the systems just didn't know, and couldn't work out, enough about the world. It's not good enough to know wonderfully about just some particular domain; you have to cover enough domains at enough depth to achieve common sense about the linguistic forms you see.

I always conceived Wolfram|Alpha as a kind of all-encompassing project. And what's now clear is that to succeed it's got to be that way. Solving a part of the problem is not enough.

The fact that as of today we've reached a 90% success rate in query understanding is a remarkable achievement—that shows we're definitely on the right track. And indeed, looking at the Wolfram|Alpha query stream, in many domains we're definitely at least on a par with typical human query-understanding performance. We're not in the running for the Turing test, though: Wolfram|Alpha doesn't currently do conversational exchanges, but more important, it knows and can compute far too much to pass for a human.

And indeed after all these years perhaps it's time to upgrade the Turing test, recognizing that computers should actually be able to do much more than humans. And from the point of view of user experience, probably the single most obvious metric is the banishment of artificial stupidity.

When Wolfram|Alpha was first released, it was quite common to run into artificial stupidity even in casual use. And I for one had no idea how long it would take to overcome it. But now, just three years later, I am quite pleased at how far we've got. It's certainly still possible to find artificial stupidity in Wolfram|Alpha (and it's quite fun to try). But it's definitely more difficult.

With all the knowledge and computation that we've put into Wolfram|Alpha, we're successfully making it not only smarter but also less stupid. And we're continuing to progress down the exponential curve toward perfect query understanding.

Scientific Bug Hunting in the Cloud: An Unexpected CEO Adventure

April 16, 2015

The Wolfram Cloud Needs to Be Perfect

The Wolfram Cloud is coming out of beta soon (yay!), and right now I'm spending much of my time working to make it as good as possible (and, by the way, it's getting to be really great!). Mostly I concentrate on defining high-level function and strategy. But I like to understand things at every level, and as a CEO, one's ultimately responsible for everything. And at the beginning of March I found myself diving deep into something I never expected...

Here's the story. As a serious production system that lots of people will use to do things like run businesses, the Wolfram Cloud should be as fast as possible. Our metrics were saying that typical speeds were good, but subjectively when I used it something felt wrong. Sometimes it was plenty fast, but sometimes it seemed way too slow.

We've got excellent software engineers, but months were going by, and things didn't seem to be changing. Meanwhile, we'd just released the Wolfram Data Drop. So I thought, why don't I just run some tests myself, maybe collecting data in our nice new Wolfram Data Drop?

A great thing about the Wolfram Language is how friendly it is for busy people: even if you only have time to dash off a few lines of code, you can get real things done. And in this case, I only had to run three lines of code to find a problem.

First, I deployed a web API for a trivial Wolfram Language program to the Wolfram Cloud:

```
In[ ]:= CloudDeploy[APIFunction[{}, 1 &]]

Out[ ]= CloudObject[https://www.wolframcloud.com/objects/b705a652… 153d0f2f2bdf]
```

Then I called the API 50 times, measuring how long each call took (%
here stands for the previous result):

```
In[•]:= Table[First[AbsoluteTiming[URLExecute[%]]], {50}]
```

```
Out[•]= {0.216327, 0.142085, 0.139396, 0.139914, 0.152744, 0.171712, 0.134547, 0.206898, 0.134823,
    0.13257, 0.137468, 0.133294, 0.133366, 0.137417, 0.13801, 0.15432, 0.138903, 0.138343,
    0.135743, 0.133329, 0.134967, 0.137378, 0.137594, 0.136514, 0.146412, 0.133264,
    0.130663, 0.132335, 0.128346, 0.13426, 0.133264, 0.133136, 0.142861, 0.129275,
    0.132988, 0.137732, 0.133475, 0.136812, 0.133644, 0.136444, 0.172769, 0.134811,
    0.129657, 0.130973, 0.129643, 0.127972, 0.131521, 0.131099, 0.130056, 0.131849}
```

Then I plotted the sequence of times for the calls:

```
In[•]:= ListLinePlot[%]
```

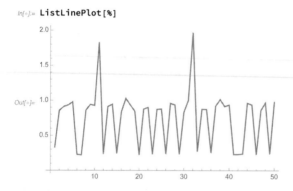

And immediately there seemed to be something crazy going on.
Sometimes the time for each call was 220 ms or so, but often it was 900
ms, or even twice that long. And the craziest thing was that the times
seemed to be quantized!

I made a histogram:

```
In[•]:= Histogram[%%, 40]
```

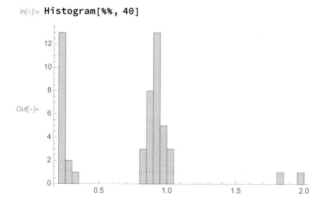

And sure enough, there were a few fast calls on the left, then a second peak of slow calls, and a third "outcropping" of very slow calls. It was weird!

I wondered whether the times were always like this. So I set up a periodic scheduled task to do a burst of API calls every few minutes, and put their times in the Wolfram Data Drop. I left this running overnight... and when I came back the next morning, this is what I saw:

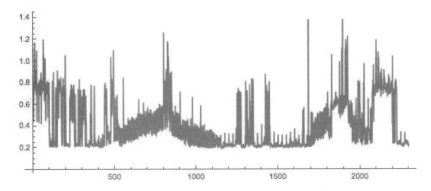

Even weirder! Why the large-scale structure? I could imagine that, for example, a particular node in the cluster might gradually slow down (not that it should), but why would it then slowly recover?

My first thought was that perhaps I was seeing network issues, given that I was calling the API on a test cloud server more than 1000 miles away. So I looked at ping times. But apart from a couple of weird spikes (hey, it's the internet!), the times were very stable.

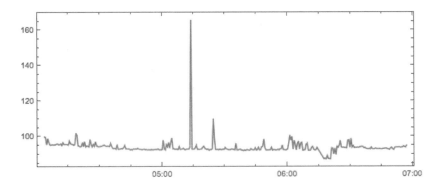

Something's Wrong inside the Servers

OK, so it must be something on the servers themselves. There's a lot of new technology in the Wolfram Cloud, but most of it is pure Wolfram Language code, which is easy to test. But there's also generic modern server infrastructure below the Wolfram Language layer. Much of this is fundamentally the same as what Wolfram|Alpha has successfully used for half a dozen years to serve billions of results, and what webMathematica started using even nearly a decade earlier. But being a more demanding computational system, the Wolfram Cloud is set up slightly differently.

And my first suspicion was that this different setup might be causing something to go wrong inside the webserver layer. Eventually I hope we'll have pure Wolfram Language infrastructure all the way down, but for now we're using a webserver system called Tomcat that's based on Java. And at first I thought that perhaps the slowdowns might be Java garbage collection. Profiling showed that there were indeed some "stop the world" garbage-collection events triggered by Tomcat, but they were rare, and were taking only milliseconds, not hundreds of milliseconds. So they weren't the explanation.

By now, though, I was hooked on finding out what the problem was. I hadn't been this deep in the trenches of system debugging for a very long time. It felt a lot like doing experimental science. And as in experimental science, it's always important to simplify what one's studying. So I cut out most of the network by operating "cloud to cloud": calling the API from within the same cluster. Then I cut out the load balancer, that dispatches requests to particular nodes in a cluster, by locking my requests to a single node (which, by the way, external users can't do unless they have a Private Cloud). But the slowdowns stayed.

So then I started collecting more detailed data. My first step was to make the API return the absolute times when it started and finished executing Wolfram Language code, and compare those to absolute times in the wrapper code that called the API. Here's what I saw:

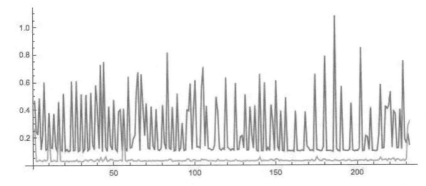

The top line shows times before the Wolfram Language code is run; the bottom line after. I collected this data in a period when the system as a whole was behaving pretty badly. And what I saw was lots of dramatic slowdowns in the "before" times—and just a few quantized slowdowns in the "after" times.

Once again, this was pretty weird. It didn't seem like the slowdowns were specifically associated with either "before" or "after." Instead, it looked more as if something was randomly hitting the system from the outside.

One confusing feature was that each node of the cluster contained (in this case) eight cores, with each core running a different instance of the Wolfram Engine. The Wolfram Engine is nice and stable, so each of these instances was running for hours to days between restarts. But I wondered if perhaps some instances might be developing problems along the way. So I instrumented the API to look at process IDs and process times, and then for example plotted total process time against components of the API call time:

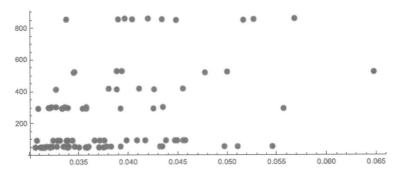

And indeed there seemed to be some tendency for "younger" processes to run API calls faster, but (particularly noting the suppressed zero on the x axis) the effect wasn't dramatic.

What's Eating the CPU?

I started to wonder about other Wolfram Cloud services running on the same machine. It didn't seem to make sense that these would lead to the kind of quantized slowdowns we were seeing, but in the interest of simplifying the system I wanted to get rid of them. At first we isolated a node on the production cluster. And then I got my very own Wolfram Private Cloud set up. Still the slowdowns were there. Though, confusingly, at different times and on different machines, their characteristics seemed to be somewhat different.

On the Private Cloud I could just log in to the raw Linux system and start looking around. The first thing I did was to read the results from the "top" and "ps axl" Unix utilities into the Wolfram Language so I could analyze them. And one thing that was immediately obvious was that lots of "system" time was being used: the Linux kernel was keeping very busy with something. And in fact, it seemed like the slowdowns might not be coming from user code at all; they might be coming from something happening in the kernel of the operating system.

So that made me want to trace system calls. I hadn't done anything like this for nearly 25 years, and my experience in the past had been that one could get lots of data, but it was hard to interpret. Now, though, I had the Wolfram Language.

Running the Linux "strace" utility while doing a few seconds of API calls gave 28,221,878 lines of output. But it took just a couple of lines of Wolfram Language code to knit together start and end times of particular system calls, and to start generating histograms of system-call durations. Doing this for just a few system calls gave me this:

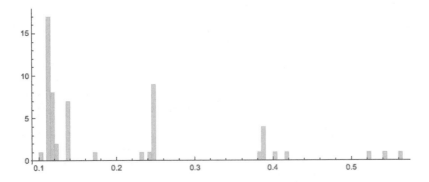

Interestingly, this showed evidence of discrete peaks. And when I looked at the system calls in these peaks they all seemed to be "futex" calls—part of the Linux thread synchronization system. So then I picked out only futex calls, and, sure enough, saw sharp timing peaks—at 250 ms, 500 ms, and 1 second:

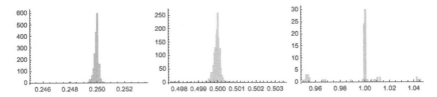

But were these really a problem? Futex calls are essentially just "sleeps"; they don't burn processor time. And actually it's pretty normal to see calls like this that are waiting for I/O to complete and so on. So to me the most interesting observation was actually that there weren't other system calls that were taking hundreds of milliseconds.

The OS Is Freezing!

So... what was going on? I started looking at what was happening on different cores of each node. Now, Tomcat and other parts of our infra- structure stack are all nicely multithreaded. Yet it seemed that whatever was causing the slowdown was freezing all the cores, even though they were running different threads. And the only thing that could do that is the operating system kernel.

But what would make a Linux kernel freeze like that? I wondered about the scheduler. I couldn't really see why our situation would lead to craziness in a scheduler. But we looked at it anyway, and tried changing a bunch of settings. No effect.

Then I had a more bizarre thought. The instances of the Wolfram Cloud I was using were running in virtual machines. What if the slow-down came from "outside the Matrix"? I asked for a version of the Wolfram Cloud running on bare metal, with no VM. But before that was configured, I found a utility to measure the "steal time" taken by the VM itself—and it was negligible.

By this point, I'd been spending an hour or two each day for several days on all of this. And it was time for me to leave for an intense trip to SXSW. Still, people in our cloud-software engineering team were revved up, and I left the problem in their capable hands.

By the time my flight arrived there was already another interesting piece of data. We'd divided each API call into 15 substeps. Then one of our physics-PhD engineers had compared the probability for a slow-down in a particular substep (on the left) to the median time spent in that substep (on the right):

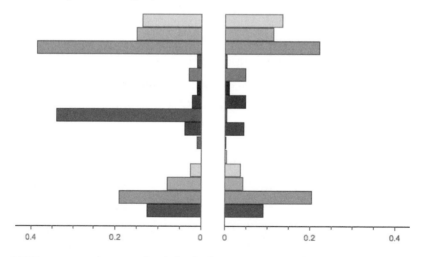

With one exception (which had a known cause), there was a good cor-relation. It really looked as if the Linux kernel (and everything running under it) was being hit by something at completely random times, causing a "slowdown event" if it happened to coincide with the running of some part of an API call.

So then the hunt was on for what could be doing this. The next suspicious thing noticed was a large amount of I/O activity. In the configuration we were testing, the Wolfram Cloud was using the NFS network file system to access files. We tried tuning NFS, changing parameters, going to asynchronous mode, using UDP instead of TCP, changing the NFS server I/O scheduler, etc. Nothing made a difference. We tried using a completely different distributed file system called Ceph. Same problem. Then we tried using local disk storage. Finally this seemed to have an effect—removing most, but not all, of the slowdown.

We took this as a clue, and started investigating more about I/O. One experiment involved editing a huge notebook on a node, while running lots of API calls to the same node:

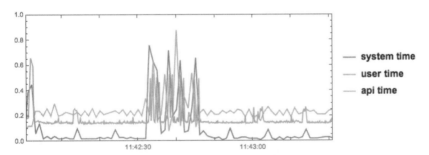

The result was interesting. During the period when the notebook was being edited (and continually saved), the API times suddenly jumped from around 100 ms to 500 ms. But why would simple file operations have such an effect on all eight cores of the node?

The Culprit Is Found

We started investigating more, and soon discovered that what seemed like "simple file operations" weren't—and we quickly figured out why. You see, perhaps five years before, early in the development of the Wolfram Cloud, we wanted to experiment with file versioning. And as a proof of concept, someone had inserted a simple versioning system named RCS.

Plenty of software systems out there in the world still use RCS, even though it hasn't been substantially updated in nearly 30 years and by now there are much better approaches (like the ones we use for infinite undo in notebooks). But somehow the RCS "proof of concept" had never been replaced in our Wolfram Cloud codebase—and it was still running on every file!

One feature of RCS is that when a file is modified even a tiny bit, lots of data (even several times the size of the file itself) ends up getting written to disk. We hadn't been sure how much I/O activity to expect in general. But it was clear that RCS was making it needlessly more intense.

Could I/O activity really hang up the whole Linux kernel? Maybe there's some mysterious global lock. Maybe the disk subsystem freezes because it doesn't flush filled buffers quickly enough. Maybe the kernel is busy remapping pages to try to make bigger chunks of memory available. But whatever might be going on, the obvious thing was just to try taking out RCS, and seeing what happened.

And so we did that, and lo and behold, the horrible slowdowns immediately went away!

So, after a week of intense debugging, we had a solution to our problem. And repeating my original experiment, everything now ran cleanly, with API times completely dominated by network transmission to the test cluster:

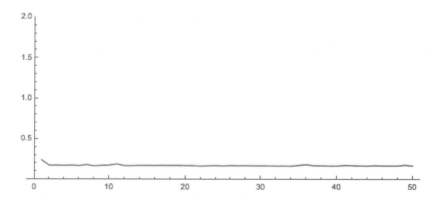

The Wolfram Language and the Cloud

What did I learn from all this? First, it reinforced my impression that the cloud is the most difficult—even hostile—development and debugging environment that I've seen in all my years in software. But second, it made me realize how valuable the Wolfram Language is as a kind of metasystem, for analyzing, visualizing, and organizing what's going on inside complex infrastructure like the cloud.

When it comes to debugging, I myself have been rather spoiled for years—because I do essentially all my programming in the Wolfram Language, where debugging is particularly easy, and it's rare for a bug to take me more than a few minutes to find. Why is debugging so easy in the Wolfram Language? I think, first and foremost, it's because the code tends to be short and readable. One also typically writes it in notebooks, where one can test out, and document, each piece of a program as one builds it up. Also critical is that the Wolfram Language is symbolic, so one can always pull out any piece of a program, and it will run on its own.

Debugging at lower levels of the software stack is a very different experience. It's much more like medical diagnosis, where one's also dealing with a complex multicomponent system, and trying to figure out what's going on from a few measurements or experiments. (I guess our versioning problem might be the analog of some horrible defect in DNA replication.)

My whole adventure in the cloud also very much emphasizes the value we're adding with the Wolfram Cloud. Because part of what the Wolfram Cloud is all about is insulating people from the messy issues of cloud infrastructure, and letting them instead implement and deploy whatever they want directly in the Wolfram Language.

Of course, to make that possible, we ourselves have needed to build all the automated infrastructure. And now, thanks to this little adventure in "scientific debugging," we're one step closer to finishing that. And indeed, as of today, the Wolfram Cloud has its APIs consistently running without any mysterious quantized slowdowns—and is rapidly approaching the point when it can move out of beta and into full production.

The Practical Business of Ontology: A Tale from the Front Lines

July 19, 2017

The Philosophy of Chemicals

"We've just got to decide: is a chemical like a city or like a number?" I spent my day yesterday—as I have for much of the past 30 years—designing new features of the Wolfram Language. And yesterday afternoon one of my meetings was a fast-paced discussion about how to extend the chemistry capabilities of the language.

At some level the problem we were discussing was quintessentially practical. But as so often turns out to be the case for things we do, it ultimately involves some deep intellectual issues. And to actually get the right answer—and to successfully design language features that will stand the test of time—we needed to plumb those depths, and talk about things that usually wouldn't be considered outside of some kind of philosophy seminar.

Part of the issue, of course, is that we're dealing with things that haven't really ever come up before. Traditional computer languages don't try to talk directly about things like chemicals; they just deal with abstract data. But in the Wolfram Language we're trying to build in as much knowledge about everything as possible, and that means we have to deal with actual things in the world, like chemicals.

We've built a whole system in the Wolfram Language for handling what we call *entities*. An entity could be a city (like New York City), or a movie, or a planet—or a zillion other things. An entity has some kind of name ("New York City"). And it has definite properties (like population, land area, founding date, ...).

We've long had a notion of chemical entities—like water, or ethanol, or tungsten carbide. Each of these chemical entities has properties, like molecular mass, or structure graph, or boiling point.

And we've got many hundreds of thousands of chemicals where we know lots of properties. But all of these are in a sense concrete chemicals: specific compounds that we could put in a test tube and do things with.

But what we were trying to figure out yesterday is how to handle abstract chemicals—chemicals that we just abstractly construct, say by giving an abstract graph representing their chemical structures. Should these be represented by entities, like water or New York City? Or should they be considered more abstract, like lists of numbers, or, for that matter, mathematical graphs?

Well, of course, among the abstract chemicals we can construct are chemicals that we already represent by entities, like sucrose or aspirin or whatever. But here there's an immediate distinction to make. Are we talking about individual molecules of sucrose or aspirin? Or about these things as bulk materials?

At some level it's a confusing distinction. Because, we might think, once we know the molecular structure, we know everything—it's just a matter of calculating it out. And some properties—like molar mass—are basically trivial to calculate from the molecular structure. But others— like melting point—are very far from trivial.

OK, but is this just a temporary problem that one shouldn't base a long-term language design on? Or is it something more fundamental that will never change? Well, conveniently enough, I happen to have done a bunch of basic science that essentially answers this: and, yes, it's something fundamental. It's connected to what I call computational irreducibility. And for example, the precise value of, say, the melting point for an infinite amount of some material may actually be fundamentally uncomputable. (It's related to the undecidability of the tiling problem; fitting in tiles is like seeing how molecules will arrange to make a solid.)

So by knowing this piece of (rather leading-edge) basic science, we know that we can meaningfully make a distinction between bulk versions of chemicals and individual molecules. Clearly there's a close relation between, say, water molecules and bulk water. But there's still something fundamentally and irreducibly different about them, and about the properties we can compute for them.

At Least the Atoms Should Be OK

Alright, so let's talk about individual molecules. Obviously they're made of atoms. And it seems like at least when we talk about atoms, we're on fairly solid ground. It might be reasonable to say that any given molecule always has some definite collection of atoms in it—though maybe we'll want to consider "parametrized molecules" when we talk about polymers and the like.

But at least it seems safe to consider types of atoms as entities. After all, each type of atom corresponds to a chemical element, and there are only a limited number of those on the periodic table. Now of course in principle one can imagine additional "chemical elements"; one could even think of a neutron star as being like a giant atomic nucleus. But again, there's a reasonable distinction to be made: almost certainly there are only a limited number of fundamentally stable types of atoms—and most of the others have ridiculously short lifetimes.

There's an immediate footnote, however. A "chemical element" isn't quite as definite a thing as one might imagine. Because it's always a mixture of different isotopes. And, say, from one tungsten mine to another, that mixture might change, giving a different effective atomic mass.

And actually this is a good reason to represent types of atoms by entities. Because then one just has to have a single entity representing tungsten that one can use in talking about molecules. And only if one wants to get properties of that type of atom that depend on qualifiers like which mine it's from does one have to deal with such things.

In a few cases (think heavy water, for example), one will need to explicitly talk about isotopes in what is essentially a chemical context. But most of the time, it's going to be enough just to specify a chemical element.

To specify a chemical element you just have to give its atomic number Z. And then textbooks will tell you that to specify a particular isotope you just have to say how many neutrons it contains. But that ignores the unexpected case of tantalum. Because, you see, one of the naturally occurring forms of tantalum (^{180m}Ta) is actually an excited state of the tantalum nucleus, which happens to be very stable. And to properly specify this, you have to give its excitation level as well as its neutron count.

In a sense, though, quantum mechanics saves one here. Because while there are an infinite number of possible excited states of a nucleus, quantum mechanics says that all of them can be characterized just by two discrete values: spin and parity.

Every isotope—and every excited state—is different, and has its own particular properties. But the world of possible isotopes is much more orderly than, say, the world of possible animals. Because quantum mechanics says that everything in the world of isotopes can be characterized just by a limited set of discrete quantum numbers.

We've gone from molecules to atoms to nuclei, so why not talk about particles too? Well, it's a bigger can of worms. Yes, there are the well-known particles like electrons and protons that are pretty easy to talk about—and are readily represented by entities in the Wolfram Language. But then there's a zoo of other particles. Some of them— just like nuclei—are pretty easy to characterize. You can basically say things like: "it's a particular excited state of a charm-quark-anti-charm-quark system" or some such. But in particle physics one's dealing with quantum field theory, not just quantum mechanics. And one can't just "count elementary particles"; one also has to deal with the possibility of virtual particles and so on. And in the end the question of what kinds of particles can exist is a very complicated one—rife with computational irreducibility. (For example, what stable states there can be of the gluon field is a much more elaborate version of something like the tiling problem I mentioned in connection with melting points.)

Maybe one day we'll have a complete theory of fundamental physics. And maybe it'll even be simple. But exciting as that will be, it's not going to help much here. Because computational irreducibility means that there's essentially an irreducible distance between what's underneath, and what phenomena emerge.

And in creating a language to describe the world, we need to talk in terms of things that can actually be observed and computed about. We need to pay attention to the basic physics—not least so we can avoid setups that will lead to confusion later. But we also need to pay attention to the actual history of science, and actual things that have been measured. Yes, there are, for example, an infinite number of possible isotopes. But for an awful lot of purposes it's perfectly useful just to set up entities for ones that are known.

The Space of Possible Chemicals

But is it the same in chemistry? In nuclear physics, we think we know all the reasonably stable isotopes that exist—so any additional and exotic ones will be very short-lived, and therefore probably not important in practical nuclear processes. But it's a different story in chemistry. There are tens of millions of chemicals that people have studied (and, for example, put into papers or patents). And there's really no limit on the molecules that one might want to consider, and that might be useful.

But, OK, so how can we refer to all these potential molecules? Well, in a first approximation we can specify their chemical structures, by giving graphs in which every node is an atom, and every edge is a bond.

What really is a "bond"? While it's incredibly useful in practical chemistry, it's at some level a mushy concept—some kind of semiclassical approximation to a full quantum mechanical story. There are some standard extra bits: double bonds, ionization states, etc. But in practice chemistry is very successfully done just by characterizing molecular structures by appropriately labeled graphs of atoms and bonds.

OK, but should chemicals be represented by entities, or by abstract graphs? Well, if it's a chemical one's already heard of, like carbon dioxide, an entity seems convenient. But what if it's a new chemical that's never been discussed before? Well, one could think about inventing a new entity to represent it.

Any self-respecting entity, though, better have a name. So what would the name be? Well, in the Wolfram Language, it could just be the graph that represents the structure. But maybe one wants something that seems more like an ordinary textual name—a string. Well, there's always the IUPAC way of naming chemicals with names like 1,1'-{[3-(dimethylamino)propyl] imino}bis-2-propanol. Or there's the more computer-friendly SMILES version: CC(CN(CCCN(C)C)CC(C)O)O. And whatever underlying graph one has, one can always generate one of these strings to represent it.

There's an immediate problem, though: the string isn't unique. In fact, however one chooses to write down the graph, it can't always be unique. A particular chemical structure corresponds to a particular graph. But there can be many ways to draw the graph—and many different representations for it. And in fact even the ("graph isomorphism") problem of determining whether two representations correspond to the same graph can be difficult to solve.

What Is a Chemical in the End?

OK, so let's imagine we represent a chemical structure by a graph. At first, it's an abstract thing. There are atoms as nodes in the graph, but we don't know how they'd be arranged in an actual molecule (and e.g. how many angstroms apart they'd be). Of course, the answer isn't completely well defined. Are we talking about the lowest-energy configuration of the molecule? (What if there are multiple configurations of the same energy?) Is the molecule supposed to be on its own, or in water, or whatever? How was the molecule supposed to have been made? (Maybe it's a protein that folded a particular way when it came off the ribosome.)

Well, if we just had an entity representing, say, "naturally occurring hemoglobin," maybe we'd be better off. Because in a sense that entity could encapsulate all these details.

But if we want to talk about chemicals that have never actually been synthesized it's a bit of a different story. And it feels as if we'd be better off just with an abstract representation of any possible chemical.

Let's talk about some other cases, and analogies. Maybe we should just treat everything as an entity. Like every integer could be an entity. Yes, there are an infinite number of them. But at least it's clear what names they should be given. With real numbers, things are already messier. For example, there's no longer the same kind of uniqueness as with integers: 0.99999... is really the same as 1.00000..., but it's written differently.

What about sequences of integers, or, for that matter, mathematical formulas? Well, every possible sequence or every possible formula could conceivably be a different entity. But this wouldn't be particularly useful, because much of what one wants to do with sequences or formulas is to go inside them, and transform their structure. But what's convenient about entities is that they're each just "single things" that one doesn't have to "go inside."

So what's the story with "abstract chemicals"? It's going to be a mixture. But certainly one's going to want to "go inside" and transform the structure. Which argues for representing the chemical by a graph.

But then there's potentially a nasty discontinuity. We've got the entity of carbon dioxide, which we already know lots of properties about. And then we've got this graph that abstractly represents the carbon dioxide molecule.

We might worry that this would be confusing both to humans and programs. But the first thing to realize is that we can distinguish what these two things are representing. The entity represents the bulk naturally occurring version of the chemical—whose properties have potentially been measured. The graph represents an abstract theoretical chemical, whose properties would have to be computed.

But obviously there's got to be a bridge. Given a concrete chemical entity, one of the properties will be the graph that represents the structure of the molecule. And given a graph, one will need some kind of ChemicalIdentify function, that—a bit like GeoIdentify or maybe ImageIdentify—tries to identify from the graph what chemical entity (if any) has a molecular structure that corresponds to that graph.

Philosophy Meets Chemistry Meets Math Meets Physics...

As I write out some of the issues, I realize how complicated all this may seem. And, yes, it is complicated. But in our meeting yesterday, it all went very quickly. Of course it helps that everyone there had seen similar issues before: this is the kind of thing that's all over the foundations of what we do. But each case is different.

And somehow this case got a bit deeper and more philosophical than usual. "Let's talk about naming stars," someone said. Obviously there are nearby stars that we have explicit names for. And some other stars may have been identified in large-scale sky surveys, and given identifiers of some kind. But there are lots of stars in distant galaxies that will never have been named. So how should we represent them?

That led to talking about cities. Yes, there are definite, chartered cities that have officially been assigned names—and we probably have essentially all of these right now in the Wolfram Language, updated regularly. But what about some village that's created for a single season by some nomadic people? How should we represent it? Well, it has a certain location, at least for a while. But is it even a definite single thing, or might it, say, devolve into two villages, or not a village at all?

One can argue almost endlessly about identity—and even existence—for many of these things. But ultimately it's not the philosophy of such things that we're interested in: we're trying to build software that people will find useful. And so what matters in the end is what's going to be useful.

Now of course that's not a precise thing to know. But it's like for language design in general: think of everything people might want to do, then see how to set up primitives that will let people do those things. Does one want some chemicals represented by entities? Yes, that's useful. Does one want a way to represent arbitrary chemical structures by graphs? Yes, that's useful.

But to see what to actually do, one has to understand quite deeply what's really being represented in each case, and how everything is related. And that's where the philosophy has to meet the chemistry, and the math, and the physics, and so on.

I'm happy to say that by the end of our hour-long meeting yesterday (informed by about 40 years of relevant experience I've had, and collectively 100+ years from people in the meeting), I think we'd come up with the essence of a really nice way to handle chemicals and chemical structures. It's going to be a while before it's all fully worked out and implemented in the Wolfram Language. But the ideas are going to help inform the way we compute and reason about chemistry for many years to come. And for me, figuring out things like this is an extremely satisfying way to spend my time. And I'm just glad that in my long-running effort to advance the Wolfram Language I get to do so much of it.

The Poetry of Function Naming

October 18, 2010

For nearly a quarter of a century, one of the responsibilities that I've taken most seriously is the shepherding of the design of Mathematica. Partly that has involved establishing foundational principles, and maintaining unity and consistency across the system. But at some point all the capabilities of Mathematica must get expressed in the individual built-in functions—like Table or NestList—that ultimately make up the system.

Each one of those functions encapsulates some piece of repeated computational work—often implemented by some deep tower of algorithms. And each one of those now 3000 or so functions requires a name.

We're currently in the closing weeks of a (spectacular!) new version of Mathematica, and I spent part of last week doing final design reviews for some fascinating new areas of the system. And as part of those design reviews, we were confirming and tweaking some of the names we're going to use for new functions.

The naming of functions is a strange and difficult art—a bit like an ultimately abstracted form of poetry. The goal is to take the concept and functionality of a function, and capture the essence of it in one, or two, or perhaps three words (like Riffle, or DeleteCases, or FixedPointList)— chosen so that when someone sees those words, they immediately get the right idea about the function. In even the most succinct forms of ordinary poetry, you get at least a handful of words to communicate with. In function names, you typically get at most perhaps three.

With enough experience, it can sometimes be pretty easy to come up with that little gem of a name for a function. Sometimes it can even seem quite obvious as soon as one thinks about the function. But sometimes it can take immense amounts of time—wrestling with what can seem like an insoluble problem of packing everything one needs to say about a function into that one little name.

It's an unforgiving and humbling activity. And the issue is almost always the same. The reason you can't find a good name is because you don't really understand with complete and ultimate clarity what the function does.

And sometimes that's because the function really isn't designed quite right. There's something muddled about it, that has to be unmuddled before you'll ever be able to find a good name.

It's very satisfying, though, when you finally crack it. These days I'm usually working on design reviews with teams of people. And when we finally get the right name, everyone on the call (yes, it's essentially always a phone call) immediately says "Oh yes, that's it." And we all feel a little stupid that we just spent an hour, or however long, just coming up with one or two words.

In ordinary human languages, new words typically develop by some form of natural selection. Usually a word will be introduced—perhaps at first as a phrase—by one person. And then it spreads, sometimes changing a bit, and either becomes popular enough to be widely understood and useful for general communication, or disappears.

But for a computer language the pattern is necessarily different. For once a function name—that corresponds to a "word" in the language—has been introduced, it must immediately be a full, permanent element of the language. For programs will be written that contain that name, and they would all have to be found and updated if that name was changed. And indeed, in Mathematica, I am proud to say that in nearly a quarter of a century, very very few names have ever had to be changed—so that a program written for Mathematica 1.0 in 1988 can still be understood and executed today by the very latest version of Mathematica.

There is also another difference between words in human languages and function names in a computer language. In a human language, there is no ultimate, absolute meaning defined for most words. Instead, the best we can do is—like in a dictionary—define words by relating them to other words.

But in a computer language, each function name ultimately refers to a particular piece of functionality that is defined in an absolute way, and can be implemented by a specific precise program.

This doesn't usually make it any easier to come up with function names, though. It just means that there's a clearer notion of the "right name": the name where a human has the best chance of correctly figuring out from it what the function does.

Function names are in a sense ultimate points of human-machine communication. They're the places where all that internal computational activity has to be connected with something that humans can understand. When the functionality is simple there are pictorial and other alternatives. But when the functionality is diverse or sophisticated we don't know any possibility other than to use language—and the linguistic construct of names for things.

The function names in Mathematica are ultimately based on English, and for the most part, they consist of ordinary English words. In ordinary natural human languages, it is possible to introduce a completely new word, and have it gradually gain popularity and understanding. But in the dynamics of computer languages—with their necessarily sudden introduction of new names—one has no choice but to leverage on people's existing understanding of a human language, like English.

Still, when we come up with function names in Mathematica today, they are in a sense not based just on "raw English." They also rely on the web of meaning that has developed through the several thousand other functions that already exist in Mathematica.

There are definite conventions about what particular kinds of names mean. (Functions that end in List generate lists; functions that begin with Image operate on images; functions that begin with Find involve some kind of searching; and so on.) There are ways that names tend to appear together in typical usage of the language. And there are definite conceptual frameworks—and metaphors—that have developed in the language and the system. (Nest refers to repeated function application; Flat refers to flattening of nested structures; Dynamic refers to dynamic interactivity; and so on.)

In ordinary human language, natural selection no doubt often favors words that follow certain patterns. Sometimes for example consistency may make words easier to remember; sometimes inconsistency makes them stand out more, and thereby easier to remember. But there is no grand plan to organize the words in a particular way: say to avoid having obscure meanings "take up" short words, or to make words easier to sort in a particular way.

But when we introduce function names in Mathematica we have both the ability—and, I think, the responsibility—to design everything. Of course, the development of Mathematica is incremental, and at any given time we can only foresee a certain amount of what will follow. But still, I take great pains to name every new function in the best possible way.

What are some of the criteria?

First, one must leverage on peoples' existing knowledge and understanding. If there is a familiar name that's already widely used, then if at all possible one must use it.

Of course, sometimes that name may only be familiar in some particular area. And it may be very short—perhaps a single letter—and incomprehensible without further context. And in that case, what we typically do in Mathematica is to burn into the name some kind of stylized context. (So, for example, the Fresnel integral S(x) has the name FresnelS.)

In building Mathematica, we've had the longstanding principle of always trying to make every function as general as possible—so that it is applicable to as wide a range of situations as possible. Sometimes, though, a function will have one particular, familiar use. But if the name of the function reflects only that use, one is shortchanging the function. For without a more general name, people will never think to apply it in other cases. (So, for example, it's List, not "vector", and it's Outer, not "outer product".)

And indeed, one of the responsibilities of function naming is that it is the names of functions that to a large extent directly determine how people will think about a function. If they are led in a particular direction by the name, that will be the direction in which they will go in using the function.

And even the very "texture" of the name is important in getting people to think correctly about functions. A sophisticated function should have a sophisticated name (like DynamicModule or EventHandler). A straight-forward, common function should have a simple name (like Length or Total). A function that does a clear but unusual thing should have an unexpected name (like Thread or Through).

By now in Mathematica there are a great many precedents for how functions should be named. And we always try to follow these precedents whenever possible. First, because they often represent good solutions to the naming problems we're now trying to solve. And second, because by following them one is maintaining a certain consistency that makes it easier for the system to grow, and for people to learn the system—and to guess about functionality they do not already know.

When one finds a good name for a function, one of the things that happens is that when people hear the name, they can successfully "unpack" it into a one-sentence description of what the function must do—often in effect just by using the name of the function as the main part of a sentence. And indeed, when we're stuck in trying to find a good name for a function, I'll often suggest that we try to write a sentence that describes what the function does—that we can perhaps use in the Documentation Center for the function, but then condense down into the nugget we need for the name itself.

One of the painful aspects of function naming is that however clever you are about it, it can never be perfect. I often claim that the only language that is perfectly consistent is the one that does nothing. As soon as there is actual functionality to represent, there are inevitably awkward cases and corners. For example, one wants to maintain consistent simplicity in naming in each area of the system. But then at the overlaps between these areas there are inconsistencies.

And sometimes one runs into limitations of English: there just isn't any familiar word or phrase for a concept, perhaps because that concept is somehow new to our experience. And in such cases what one typically has to do—just like in natural language—is to come up with an analogy.

Some of the analogies and metaphors we consider start quite wild and outlandish. But eventually they become tamer—like Sow and Reap or

Throw and Catch—and an important way to extend the linguistic base for names in Mathematica.

It might be nice if English—like Mathematica—had the feature that a particular word meant only a particular thing, or at least a class of things. But that is not how it works. A single word can act as different parts of speech, and can have wildly different meanings. Usually in actual English usage, one can disambiguate by context.

But in the tiny length of a single function name, one does not have that option. And quite often that means one has to reject some wonderful word just in order to avoid a possible misunderstanding from a different way it can be used in English. (So, for example, "Live" or "Active" can't be candidates for Dynamic—they're just too easy to misunderstand.)

If one is lucky, a thesaurus (these days in Wolfram|Alpha) will give one a word that captures the same concept but avoids the potential misunderstanding. But sometimes one has to rearrange the whole structure of the name to avoid the possibility of misunderstanding.

And yet, after all those judgment calls, after all that drilling to clarify precisely what a function does, one has to come to a conclusion: one has to settle on a definite name. That will represent the function—and all the work done to implement it—well. And that will serve as a permanent handle by which people can access some piece of functionality in Mathematica.

I have no idea now how much time I have spent over the past quarter century coming up with names in Mathematica. Each one encapsulates some idea, some creative concept—frozen in a tiny clump of words. Like little poems. Thousands of them.

Buzzword Convergence: Making Sense of Quantum Neural Blockchain AI

April 1, 2018

Not Entirely Fooling Around

What happens if you take four of today's most popular buzzwords and string them together? Does the result mean anything? Given that today is April 1 (as well as being Easter Sunday), I thought it'd be fun to explore this. Think of it as an Easter egg... from which something interesting just might hatch. And to make it clear: while I'm fooling around in stringing the buzzwords together, the details of what I'll say here are perfectly real.

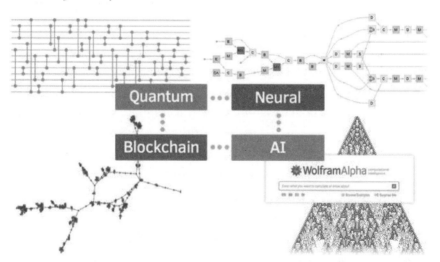

But before we can really launch into talking about the whole string of buzzwords, let's discuss some of the background to each of the buzzwords on their own.

"Quantum"

Saying something is "quantum" sounds very modern. But actually, quantum mechanics is a century old. And over the course of the past century, it's been central to understanding and calculating lots of things in the physical sciences. But even after a century, "truly quantum" technology hasn't arrived. Yes, there are things like lasers and MRIs and atomic force microscopes that rely on quantum phenomena, and needed quantum mechanics in order to be invented. But when it comes to the practice of engineering, what's done is still basically all firmly classical, with nothing quantum about it.

Today, though, there's a lot of talk about quantum computing, and how it might change everything. I actually worked on quantum computing back in the early 1980s (so, yes, it's not that recent an idea). And I have to say, I was always a bit skeptical about whether it could ever really work—or whether any "quantum gains" one might get would be counterbalanced by inefficiencies in measuring what was going on.

But in any case, in the past 20 years or so there's been all sorts of nice theoretical work on formulating the idea of quantum circuits and quantum computing. Lots of things have been done with the Wolfram Language, including an ongoing project of ours to produce a definitive symbolic way of representing quantum computations. But so far, all we can ever do is calculate about quantum computations, because the Wolfram Language itself just runs on ordinary, classical computers.

There are companies that have built what they say are (small) true quantum computers. And actually, we've been hoping to hook the Wolfram Language up to them, so we can implement a QuantumEvaluate function. But so far, this hasn't happened. So I can't really vouch for what QuantumEvaluate will (or will not) do.

But the big idea is basically this. In ordinary classical physics, one can pretty much say that definite things happen in the world. A billiard ball goes in this direction, or that. But in any particular case, it's a definite direction. In quantum mechanics, though, the idea is that an electron, say, doesn't intrinsically go in a particular, definite direction. Instead, it essentially goes in all possible directions, each with a particular amplitude. And it's only when you insist on measuring where it went that you'll get a definite answer. And if you do many measurements, you'll just see probabilities for it to go in each direction.

Well, what quantum computing is trying to do is somehow to make use of the "all possible directions" idea in order to in effect get lots of computations done in parallel. It's a tricky business, and there are only a few types of problems where the theory's been worked out—the most famous being integer factoring. And, yes, according to the theory, a big quantum computer should be able to factor a big integer fast enough to make today's cryptography infrastructure implode. But the only thing anyone so far even claims to have built along these lines is a tiny quantum computer—that definitely can't yet do anything terribly interesting.

But, OK, so one critical aspect of quantum mechanics is that there can be interference between different paths that, say, an electron can take. This is mathematically similar to the interference that happens in light, or even in water waves, just in classical physics. In quantum mechanics, though, there's supposed to be something much more intrinsic about the interference, leading to the phenomenon of entanglement, in which one basically can't ever "see the wave that's interfering"—only the effect.

In computing, though, we're not making use of any kind of interference yet. Because (at least in modern times) we're always trying to deal with discrete bits—while the typical phenomenon of interference (say in light) basically involves continuous numbers. And my personal guess is that optical computing—which will surely come—will succeed in delivering some spectacular speedups. It won't be truly "quantum", though (though it might be marketed like that). (For the technically minded, it's a complicated question how computation-theoretic results apply to continuous processes like interference-based computing.)

"Neural"

A decade ago computers didn't have any systematic way to tell whether a picture was of an elephant or a teacup. But in the past five years, thanks to neural networks, this has basically become easy. (Interestingly, the image identifier we made three years ago remains basically state of the art.)

So what's the big idea? Well, back in the 1940s people started thinking seriously about the brain being like an electrical machine. And this led to mathematical models of "neural networks"—which were proved to be equivalent in computational power to mathematical models of digital computers. Over the years that followed, billions of actual digital electronic computers were built. And along the way, people (including me) experimented with neural networks, but nobody could get them to do anything terribly interesting. (Though for years they were quietly used for things like optical character recognition.)

But then, starting in 2012, a lot of people suddenly got very excited, because it seemed like neural nets were finally able to do some very interesting things, at first especially in connection with images.

So what happened? Well, a neural net basically corresponds to a big mathematical function, formed by connecting together lots of smaller functions, each involving a certain number of parameters ("weights"). At the outset, the big function basically just gives random outputs. But the way the function is set up, it's possible to "train the neural net" by tuning the parameters inside it so that the function will give the outputs one wants.

It's not like ordinary programming where one explicitly defines the steps a computer should follow. Instead, the idea is just to give examples of what one wants the neural net to do, and then to expect it to interpolate between them to work out what to do for any particular input. In practice one might show a bunch of images of elephants, and a bunch of images of teacups, and then do millions of little updates to the parameters to get the network to output "elephant" when it's fed an elephant, and "teacup" when it's fed a teacup.

But here's the crucial idea: the neural net is somehow supposed to generalize from the specific examples it's shown—and it's supposed to say that anything that's "like" an elephant example is an elephant, even if its particular pixels are quite different. Or, said another way, there are lots of images that might be fed to the network that are in the "basin of attraction" for "elephant" as opposed to "teacup". In a mechanical analogy, one might say that there are lots of places water might fall on a landscape, while still ending up flowing to one lake rather than another.

At some level, any sufficiently complicated neural net can in principle be trained to do anything. But what's become clear is that for lots of practical tasks (that turn out to overlap rather well with some of what our brains seem to do easily) it's realistic with feasible amounts of GPU time to actually train neural networks with a few million elements to do useful things. And, yes, in the Wolfram Language we've now got a rather sophisticated symbolic framework for training and using neural networks— with a lot of automation (that itself uses neural nets) for everything.

"Blockchain"

The word "blockchain" was first used in connection with the invention of Bitcoin in 2008. But of course the idea of a blockchain had precursors. In its simplest form, a blockchain is like a ledger, in which successive entries are coded in a way that depends on all previous entries.

Crucial to making this work is the concept of hashing. Hashing has always been one of my favorite practical computation ideas (and I even independently came up with it when I was about 13 years old, in 1973). What hashing does is to take some piece of data, like a text string, and make a number (say between 1 and a million) out of it. It does this by "grinding up the data" using some complicated function that always gives the same result for the same input, but will almost always give different results for different inputs. There's a function called Hash in the Wolfram Language, and for example applying it to the previous paragraph of text gives 8643827914633641131.

OK, but so how does this relate to blockchain? Well, back in the 1980s people invented "cryptographic hashes" (and actually they're very related to things I've done on computational irreducibility). A cryptographic hash has the feature that while it's easy to work out the hash for a particular piece of data, it's very hard to find a piece of data that will generate a given hash.

So let's say you want to prove that you created a particular document at a particular time. Well, you could compute a hash of that document, and publish it in a newspaper (and I believe Bell Labs actually used to do this every week back in the 1980s). And then if anyone ever says "no, you didn't have that document yet" on a certain date, you can just say "but look, its hash was already in every copy of the newspaper!"

The idea of a blockchain is that one has a series of blocks, with each containing certain content, together with a hash. And then the point is that the data from which that hash is computed is a combination of the content of the block, and the hash of the preceding block. So this means that each block in effect confirms everything that came before it on the blockchain.

In cryptocurrencies like Bitcoin the big idea is to be able to validate transactions, and, for example, be able to guarantee just by looking at the blockchain that nobody has spent the same bitcoin twice.

How does one know that the blocks are added correctly, with all their hashes computed, etc.? Well, the point is that there's a whole decentralized network of thousands of computers around the world that store the blockchain, and there are lots of people (well, actually not so many in practice these days) competing to be the one to add each new block (and include transactions people have submitted that they want in it).

The rules are (more or less) that the first person to add a block gets to keep the fees offered on the transactions in it. But each block gets "confirmed" by lots of people including this block in their copy of the blockchain, and then continuing to add to the blockchain with this block in it.

In the latest version of the Wolfram Language, BlockchainBlockData[−1, BlockchainBase → "Bitcoin"] gives a symbolic representation of the latest block that we've seen be added to the Bitcoin blockchain. And by the time maybe five more blocks have been added, we can be pretty sure everyone's satisfied that the block is correct. (Yes, there's an analogy with measurement in quantum mechanics here, which I'll be talking about soon.)

Traditionally, when people keep ledgers, say of transactions, they'll have one central place where a master ledger is maintained. But with a blockchain the whole thing can be distributed, so you don't have to trust any single entity to keep the ledger correct.

And that's led to the idea that cryptocurrencies like Bitcoin can flourish without central control, governments or banks involved. And in the last couple of years there's been lots of excitement generated by people making large amounts of money speculating on cryptocurrencies.

But currencies aren't the only thing one can use blockchains for, and Ethereum pioneered the idea that in addition to transactions, one can run arbitrary computations at each node. Right now with Ethereum the results of each computation are confirmed by being run on every single computer in the network, which is incredibly inefficient. But the bigger point is just that computations can be running autonomously on the network. And the computations can interact with each other, defining "smart contracts" that run autonomously, and say what should happen in different circumstances.

Pretty much any nontrivial smart contract will eventually need to know about something in the world ("did it rain today?", "did the package arrive?", etc.), and that has to come from off the blockchain—from an "oracle". And it so happens (yes, as a result of a few decades of work) that our Wolfram Knowledgebase, which powers Wolfram|Alpha, etc., provides the only realistic foundation today for making such oracles.

"AI"

Back in the 1950s, people thought that pretty much anything human intelligence could do, it'd soon be possible to make artificial (machine) intelligence do better. Of course, this turned out to be much harder than people expected. And in fact the whole concept of "creating artificial intelligence" pretty much fell into disrepute, with almost nobody wanting to market their systems as "doing AI".

But about five years ago—particularly with the unexpected successes in neural networks—all that changed, and AI was back, and cooler than ever.

What is AI supposed to be, though? Well, in the big picture I see it as being the continuation of a long trend of automating things that humans previously had to do for themselves—and in particular doing that through computation. But what makes a computation an example of AI, and not just, well, a computation?

I've built a whole scientific and philosophical structure around something I call the Principle of Computational Equivalence, that basically says that the universe of possible computations—even done by simple systems—is full of computations that are as sophisticated as one can ever get, and certainly as our brains can do.

In doing engineering, and in building programs, though, there's been a tremendous tendency to try to prevent anything too sophisticated from happening—and to set things up so that the systems we build just follow exactly steps we can foresee. But there's much more to computation than that, and in fact I've spent much of my life building systems that make use of this.

Wolfram|Alpha is a great example. Its goal is to take as much knowledge about the world as possible, and make it computable, then to be able to answer questions as expertly as possible about it. Experientially, it "feels like AI", because you get to ask it questions in natural language like a human, then it computes answers, often with unexpected sophistication.

Most of what's inside Wolfram|Alpha doesn't work anything like brains probably do, not least because it's leveraging the last few hundred years of formalism that our civilization has developed, that allow us to be much more systematic than brains naturally are.

Some of the things modern neural nets do (and, for example, our machine learning system in the Wolfram Language does) perhaps work a little more like brains. But in practice what really seems to make things "seem like AI" is just that they're operating on the basis of sophisticated computations whose behavior we can't readily understand.

These days the way I see it is that out in the computational universe there's amazing computational power. And the issue is just to be able to harness that for useful human purposes. Yes, "an AI" can go off and do all sorts of computations that are just as sophisticated as our brains. But the issue is: can we align what it does with things we care about doing?

And, yes, I've spent a large part of my life building the Wolfram Language, whose purpose is to provide a computational communication language in which humans can express what they want in a form suitable for computation. There's lots of "AI power" out there in the computational universe; our challenge is to harness it in a way that's useful to us.

Oh, and we want to have some kind of computational smart contracts that define how we want the AIs to behave (e.g. "be nice to humans").

And, yes, I think the Wolfram Language is going to be the right way to express those things, and build up the "AI Constitutions" we want.

Common Themes

At the outset, it might seem as if "quantum", "neural", "blockchain", and "AI" are all quite separate concepts, without a lot of commonality. But actually it turns out that there are some amazing common themes.

One of the strongest has to do with complexity generation. And in fact, in their different ways, all the things we're talking about rely on complexity generation.

What do I mean by complexity generation? One day I won't have to explain this. But for now I probably still do. And somehow I find myself always showing the same picture—of my all-time favorite science discovery, the rule 30 automaton. Here it is:

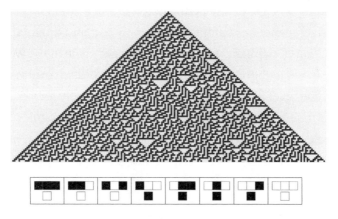

And the point here is that even though the rule (or program) is very simple, the behavior of the system just spontaneously generates complexity, and apparent randomness. And what happens is complicated enough that it shows what I call "computational irreducibility", so that you can't reduce the computational work needed to see how it will behave: you essentially just have to follow each step to find out what will happen.

There are all sorts of important phenomena that revolve around complexity generation and computational irreducibility. The most obvious is just the fact that sophisticated computation is easy to get— which is in a sense what makes something like AI possible.

But OK, how does this relate to blockchain? Well, complexity generation is what makes cryptographic hashing possible. It's what allows a simple algorithm to make enough apparent randomness to successfully be used as a cryptographic hash.

In the case of something like Bitcoin, there's another connection too: the protocol needs people to have to make some investment to be able to add blocks to the blockchain, and the way this is achieved is (bizarrely enough) by forcing them to do irreducible computations that effectively cost computer time.

What about neural nets? Well, the very simplest neural nets don't involve much complexity at all. If one drew out their "basins of attraction" for different inputs, they'd just be simple polygons. But in useful neural nets the basins of attraction are much more complicated.

It's most obvious when one gets to recurrent neural nets, but it happens in the training process for any neural net: there's a computational process that effectively generates complexity as a way to approximate things like the distinctions ("elephant" vs. "teacup") that get made in the world.

Alright, so what about quantum mechanics? Well, quantum mechanics is at some level full of randomness. It's essentially an axiom of the traditional mathematical formalism of quantum mechanics that one can only compute probabilities, and that there's no way to "see under the randomness".

I personally happen to think it's pretty likely that that's just an approximation, and that if one could get "underneath" things like space and time, we'd see how the randomness actually gets generated.

But even in the standard formalism of quantum mechanics, there's a kind of complementary place where randomness and complexity generation is important, and it's in the somewhat mysterious process of measurement.

Let's start off by talking about another phenomenon in physics: the Second Law of Thermodynamics, or Law of Entropy Increase. This law says that if you start, for example, a bunch of gas molecules in a very orderly configuration (say all in one corner of a box), then with overwhelming probability they'll soon randomize (and e.g. spread out randomly all over the box). And, yes, this kind of trend towards randomness is something we see all the time.

Here's the strange part: if we look at the laws for, say, the motion of individual gas molecules, they're completely reversible—so just as they say that the molecules can randomize themselves, so also they say that they should be able to unrandomize themselves.

But why do we never see that happen? It's always been a bit mysterious, but I think there's a clear answer, and it's related to complexity generation and computational irreducibility. The point is that when the gas molecules randomize themselves, they're effectively encrypting the initial conditions they were given.

It's not impossible to place the gas molecules so they'll unrandomize rather than randomize; it's just that to work out how to do this effectively requires breaking the encryption—or in essence doing something very much like what's involved in Bitcoin mining.

OK, so how does this relate to quantum mechanics? Well, quantum mechanics itself is fundamentally based on probability amplitudes, and interference between different things that can happen. But our experience of the world is that definite things happen. And the bridge from quantum mechanics to this involves the rather "bolted-on" idea of quantum measurement.

The notion is that some little quantum effect ("the electron ends up with spin up, rather than down") needs to get amplified to the point where one can really be sure what happened. In other words, one's measuring device has to make sure that the little quantum effect associated with one electron cascades so that it's spread across lots and lots of electrons and other things.

And here's the tricky part: if one wants to avoid interference being possible (so we can really perceive something "definite" as having happened), then one needs to have enough randomness that things can't somehow equally well go backwards—just like in thermodynamics.

So even though pure quantum circuits as one imagines them for practical quantum computers typically have a sufficiently simple mathematical structure that they (presumably) don't intrinsically generate complexity, the process of measuring what they do inevitably must generate complexity. (And, yes, it's a reasonable question whether that's in some sense where the randomness one sees "really" comes from... but that's a different story.)

Reversibility, Irreversibility, and More

Reversibility and irreversibility are a strangely common theme, at least between "quantum", "neural", and "blockchain". If one ignores measurement, a fundamental feature of quantum mechanics is that it's reversible. What this means is that if one takes a quantum system, and lets it evolve in time, then whatever comes out one will always, at least in principle, be able to take and run backwards, to precisely reproduce where one started from.

Typical computation isn't reversible like that. Consider an OR gate, that might be a basic component in a computer. In p OR q, the result will be true if either p or q is true. But just knowing that the result is "true", you can't figure out which of p and q (or both) is true. In other words, the OR operation is irreversible: it doesn't preserve enough information for you to invert it.

In quantum circuits, one uses gates that, say, take two inputs (say p and q), and give two outputs (say p' and q'). And from those two outputs one can always uniquely reproduce the two inputs.

OK, but now let's talk about neural nets. Neural nets as they're usually conceived are fundamentally irreversible. Here's why. Imagine (again) that you make a neural network to distinguish elephants and teacups. To make that work, a very large number of different possible input images all have to map, say, to "elephant". It's like the OR gate, but more so. Just knowing the result is "elephant" there's no unique way to invert the computation. And that's the whole point: one wants anything that's enough like the elephant pictures one showed to still come out as "elephant"; in other words, irreversibility is central to the whole operation of at least this kind of neural net.

So, OK, then how could one possibly make a quantum neural net? Maybe it's just not possible. But if so, then what's going on with brains? Because brains seem to work very much like neural nets. And yet brains are physical systems that presumably follow quantum mechanics. So then how are brains possible?

At some level the answer has to do with the fact that brains dissipate heat. Well, what is heat? Microscopically, heat is the random motion of things like molecules. And one way to state the Second Law of Thermodynamics (or the Law of Entropy Increase) is that under normal circumstances those random motions never spontaneously organize themselves into any kind of systematic motion. In principle all those molecules could start moving in just such a way as to turn a flywheel. But in practice nothing like that ever happens. The heat just stays as heat, and doesn't spontaneously turn into macroscopic mechanical motion.

OK, but so let's imagine that microscopic processes involving, say, collisions of molecules, are precisely reversible—as in fact they are according to quantum mechanics. Then the point is that when lots of molecules are involved, their motions can get so "encrypted" that they just seem random. If one could look at all the details, there'd still be enough information to reverse everything. But in practice one can't do that, and so it seems like whatever was going on in the system has just "turned into heat."

So then what about producing "neural net behavior"? Well, the point is that while one part of a system is, say, systematically "deciding to say elephant", the detailed information that would be needed to go back to the initial state is getting randomized, and turning into heat.

To be fair, though, this is glossing over quite a bit. And in fact I don't think anyone knows how one can actually set up a quantum system (say a quantum circuit) that behaves in this kind of way. It'd be pretty interesting to do so, because it'd potentially tell us a lot about the quantum measurement process.

To explain how one goes from quantum mechanics in which everything is just an amplitude, to our experience of the world in which definite things seem to happen, people sometimes end up trying to appeal to mystical features of consciousness. But the point about a quantum neural net is that it's quantum mechanical, yet it "comes to definite conclusions" (e.g. elephant vs. teacup).

Is there a good toy model for such a thing? I suspect one could create one from a quantum version of a cellular automaton that shows phase transition behavior—actually not unlike the detailed mechanics of a real quantum magnetic material. And what will be necessary is that the system has enough components (say spins) that the "heat" needed to compensate for its apparent irreversible behavior will stay away from the part where the irreversible behavior is observed.

Let me make a perhaps slightly confusing side remark. When people talk about "quantum computers", they are usually talking about quantum circuits that operate on qubits (quantum analog of binary bits). But sometimes they actually mean something different: they mean quantum annealing devices.

Imagine you've got a bunch of dominoes and you're trying to arrange them on the plane so that some matching condition associated with the markings on them is always satisfied. It turns out this can be a very hard problem. It's related to computational irreducibility (and perhaps to problems like integer factoring). But in the end, to find out, say, the configuration that does best in satisfying the matching condition every-where, one may effectively have to essentially just try out all possible configurations, and see which one works best.

Well, OK, but let's imagine that the dominoes were actually molecules, and the matching condition corresponds to arranging molecules to minimize energy. Then the problem of finding the best overall configuration is like the problem of finding the minimum energy configuration for the molecules, which physically should correspond to the most stable solid structure that can be formed from the molecules.

And, OK, it might be hard to compute that. But what about an actual physical system? What will the molecules in it actually do when one cools it down? If it's easy for the molecules to get to the lowest energy configuration, they'll just do it, and one will have a nice crystalline solid.

People sometimes assume that "the physics will always figure it out," and that even if the problem is computationally hard, the molecules will always find the optimal solution. But I don't think this is actually true—and I think what instead will happen is that the material will turn mushy, not quite liquid and not quite solid, at least for a long time.

Still, there's the idea that if one sets up this energy minimization problem quantum mechanically, then the physical system will be successful at finding the lowest energy state. And, yes, in quantum mechanics it might be harder to get stuck in local minima, because there is tunneling, etc.

But here's the confusing part: when one trains a neural net, one ends up having to effectively solve minimization problems like the one I've described ("which values of weights make the network minimize the error in its output relative to what one wants?"). So people end up sometimes talking about "quantum neural nets", meaning domino-like arrays which are set up to have energy minimization problems that are mathematically equivalent to the ones for neural nets.

(Yet another connection is that convolutional neural nets—of the kind used for example in image recognition—are structured very much like cellular automata, or like dynamic spin systems. But in training neural nets to handle multiscale features in images, one seems to end up with scale invariance similar to what one sees at critical points in spin systems, or their quantum analogs, as analyzed by renormalization group methods.)

OK, but let's return to our whole buzzword string. What about blockchain? Well, one of the big points about a blockchain is in a sense to be as irreversible as possible. Once something has been added to a blockchain, one wants it to be inconceivable that it should ever be reversed out.

How is that achieved? Well, it's curiously similar to how it works in thermodynamics or in quantum measurement. Imagine someone adds a block to their copy of a blockchain. Well, then the idea is that lots of other people all over the world will make their own copies of that block on their own blockchain nodes, and then go on independently adding more blocks from there.

Bad things would happen if lots of the people maintaining block-chain nodes decided to collude to not add a block, or to modify it, etc. But it's a bit like with gas molecules (or degrees of freedom in quantum measurement). By the time everything is spread out among enough different components, it's extremely unlikely that it'll all concentrate together again to have some systematic effect.

Of course, people might not be quite like gas molecules (though, frankly, their observed aggregate behavior, e.g. jostling around in a crowd, is often strikingly similar). But all sorts of things in the world seem to depend on an assumption of randomness. And indeed, that's probably necessary to maintain stability and robustness in markets where trading is happening.

OK, so when a blockchain tries to ensure that there's a "definite history", it's doing something very similar to what a quantum measurement has to do. But just to close the loop a little more, let's ask what a quantum blockchain might be like.

Yes, one could imagine using quantum computing to somehow break the cryptography in a standard blockchain. But the more interesting (and, in my view, realistic) possibility is to make the actual operation of the blockchain itself be quantum mechanical.

In a typical blockchain, there's a certain element of arbitrariness in how blocks get added, and who gets to do it. In a "proof of work" scheme (as used in Bitcoin and currently also Ethereum), to find out how to add a new block one searches for a "nonce"—a number to throw in to make a hash come out in a certain way. There are always many possible nonces (though each one is hard to find), and the typical strategy is to search randomly for them, successively testing each candidate.

But one could imagine a quantum version in which one is in effect searching in parallel for all possible nonces, and as a result producing many possible blockchains, each with a certain quantum amplitude. And to fill out the concept, imagine that—for example in the case of Ethereum—all computations done on the blockchain were reversible quantum ones (achieved, say, with a quantum version of the Ethereum Virtual Machine).

But what would one do with such a blockchain? Yes, it would be an interesting quantum system with all kinds of dynamics. But to actually connect it to the world, one has to get data on and off the blockchain—or, in other words, one has to do a measurement. And the act of that measurement would in effect force the blockchain to pick a definite history.

OK, so what about a "neural blockchain"? At least today, by far the most common strategy with neural nets is first to train them, then to put them to work. (One can train them "passively" by just feeding them a fixed set of examples, or one can train them "actively" by having them in effect "ask" for the examples they want.) But by analogy with people, neural nets can also have "lifelong learning", in which they're continually getting updated based on the "experiences" they're having.

So how do the neural nets record these experiences? Well, by changing various internal weights. And in some ways what happens is like what happens with blockchains.

Science fiction sometimes talks about direct brain-to-brain transfer of memories. And in a neural net context this might mean just taking a big block of weights from one neural net and putting it into another. And, yes, it can work well to transfer definite layers in one network to another (say to transfer information on what features of images are worth picking out). But if you try to insert a "memory" deep inside a network, it's a different story. Because the way a memory is represented in a network will depend on the whole history of the network.

It's like in a blockchain: you can't just replace one block and expect everything else to work. The whole thing has been knitted into the sequence of things that happen through time. And it's the same thing with memories in neural nets: once a memory has formed in a certain way, subsequent memories will be built on top of this one.

Bringing It Together

At the outset, one might have thought that "quantum", "neural", and "blockchain" (not to mention "AI") didn't have much in common (other than that they're current buzzwords)—and that in fact they might in some sense be incompatible. But what we've seen is that actually there are all sorts of connections between them, and all sorts of fundamental phenomena that are shared between systems based on them.

So what might a "quantum neural blockchain AI" ("QNBAI") be like?

Let's look at the pieces again. A single blockchain node is a bit like a single brain, with a definite memory. But in a sense the whole blockchain network becomes robust through all the interactions between different blockchain nodes. It's a little like how human society and human knowledge develop.

Let's say we've got a "raw AI" that can do all sorts of computation. Well, the big issue is whether we can find a way to align what it can do with things that we humans think we want to do. And to make that alignment, we essentially have to communicate with the AI at a level of abstraction that transcends the details of how it works: in effect, we have to have some symbolic language that we both understand, and that for example AI can translate into the details of how it operates.

Inside the AI it may end up using all kinds of "concepts" (say to distinguish one class of images from another). But the question is whether those concepts are ones that we humans in a sense "culturally understand." In other words, are those concepts (and, for example, the words for them) ones that there's a whole widely understood story about?

In a sense, concepts that we humans find useful for communication are ones that have been used in all sorts of interactions between different humans. The concepts become robust by being "knitted into" the thought patterns of many interacting brains, a bit like the data put on a blockchain becomes a robust part of "collective blockchain memory" through the interactions between blockchain nodes.

OK, so there's something strange here. At first it seemed like QNBAIs would have to be something completely exotic and unfamiliar (and perhaps impossible). But somehow as we go over their features they start to seem awfully familiar—and actually awfully like us.

Yup, according to the physics, we know we are "quantum." Neural nets capture many core features of how our brains seem to work. Blockchain—at least as a general concept—is somehow related to individual and societal memory. And AI, well, AI in effect tries to capture what's aligned with human goals and intelligence in the computational universe—which is also what we're doing.

OK, so what's the closest thing we know to a QNBAI? Well, it's probably all of us!

Maybe that sounds crazy. I mean, why should a string of buzzwords from 2018 connect like that? Well, at some level perhaps there's an obvious answer: we tend to create and study things that are relevant to us, and somehow revolve around us. And, more than that, the buzzwords of today are things that are somehow just within the scope that we can now think about with the concepts we've currently developed—and that are somehow connected through them.

I must say that when I chose these buzzwords I had no idea they'd connect at all. But as I've tried to work through things in writing this, it's been remarkable how much connection I've found. And, yes, in a fittingly bizarre end to a somewhat bizarre journey, it does seem to be the case that a string plucked from today's buzzword universe has landed very close to home. And maybe in the end—at least in some sense—we are our buzzwords!

Oh My Gosh, It's Covered in Rule 30s!

June 1, 2017

A British Train Station

A week ago a new train station, named "Cambridge North," opened in Cambridge, UK. Normally such an event would be far outside my sphere of awareness. (I think I last took a train to Cambridge in 1975.) But last week people started sending me pictures of the new train station, wondering if I could identify the pattern on it:

And, yes, it does indeed look a lot like patterns I've spent years studying—that come from simple programs in the computational universe. My first—and still favorite—examples of simple programs are one-dimensional cellular automata like this:

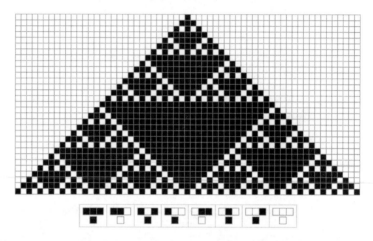

The system evolves line by line from the top, determining the color of each cell according to the rule underneath. This particular cellular automata I called "rule 182," because the bit pattern in the rule corresponds to the number 182 in binary. There are altogether 256 possible cellular automata like this, and this is what all of them do:

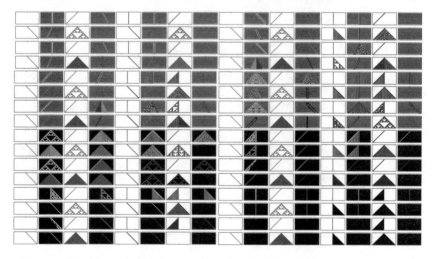

Many of them show fairly simple behavior. But the huge surprise I got when I first ran all these cellular automata in the early 1980s is that even though all the rules are very simple to state, some of them generate very complex behavior. The first in the list that does that—and still my favorite example—is rule 30:

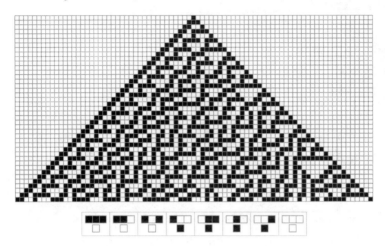

If one runs it for 400 steps one gets this:

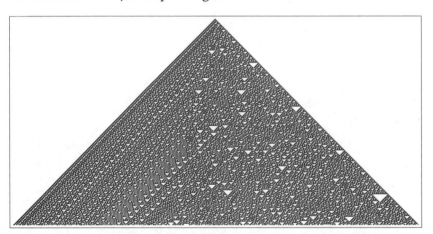

And, yes, it's remarkable that starting from one black cell at the top, and just repeatedly following a simple rule, it's possible to get all this complexity. I think it's actually an example of a hugely important phenomenon, that's central to how complexity gets made in nature, as well as to how we can get a new level of technology. And in fact, I think it's important enough that I spent more than a decade writing a 1200-page book (that just celebrated its 15th anniversary) based on it.

And for years I've actually had rule 30 on my business cards:

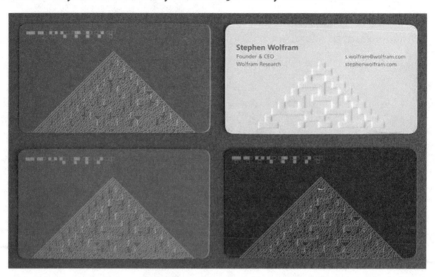

But back to the Cambridge North train station. Its pattern is obviously not completely random. But if it was made by a rule, what kind of rule? Could it be a cellular automaton?

I zoomed in on a photograph of the pattern:

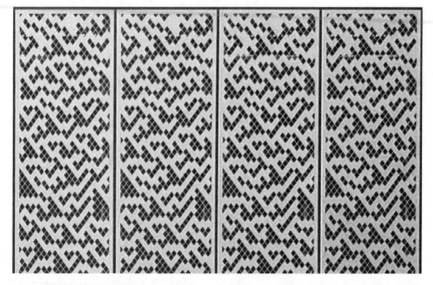

Suddenly, something seemed awfully familiar: the triangles, the stripes, the L shapes. Wait a minute... it couldn't actually be my favorite rule of all time, rule 30?

Clearly the pattern is tipped 45° from how I'd usually display a cellular automaton. And there are black triangles in the photograph, not white ones like in rule 30. But if one black-white inverts the rule (so it's now rule 135), one gets this:

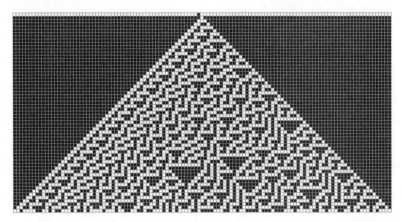

And, yes, it's the same kind of pattern as in the photograph! But if it's rule 30 (or rule 135) what's its initial condition? Rule 30 can actually be used as a cryptosystem—because it can be hard (maybe even NP complete) to reconstruct its initial condition.

But, OK, if it's my favorite rule, I wondered if maybe it's also my favorite initial condition—a single black cell. And, yes, it is! The train station pattern comes exactly from the (inverted) right-hand edge of my favorite rule 30 pattern!

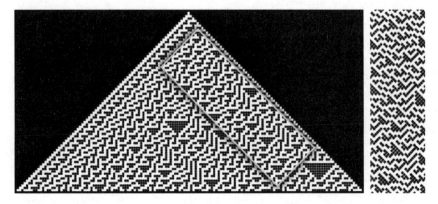

Here's the Wolfram Language code. First run the cellular automaton, then rotate the pattern:

In[ ]:= `Rotate[ArrayPlot[CellularAutomaton[135, {{1}, 0}, 40], Mesh → True], -45 °]`

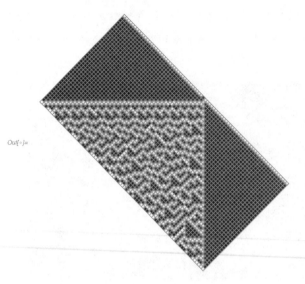

Out[ ]=

It's a little trickier to pull out precisely the section of the pattern that's used. Here's the code (the PlotRange is what determines the part of the pattern that's shown):

In[ ]:= `Graphics[Rotate[First[ArrayPlot[CellularAutomaton[135, {{1}, 0}, 80],`
`        Mesh → True]], -45 °], PlotRange → {{83, 104}, {-12, 60}}]`

Out[ ]=

OK, so where is this pattern actually used at the train station? Everywhere!

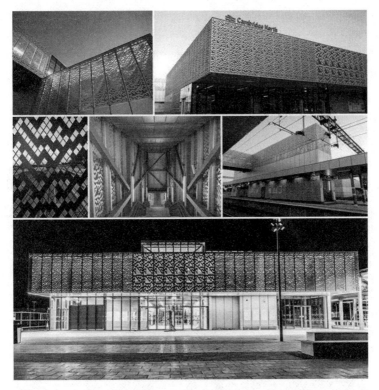

It's made of perforated aluminum. You can actually look through it, reminiscent of an old latticed window. From inside, the pattern is left-right reversed—so if it's rule 135 from outside, it's rule 149 from inside. And at night, the pattern is black-white inverted, because there's light coming from inside—so from the outside it's "rule 135 by day, and rule 30 at night."

What are some facts about the rule 30 pattern? It's extremely hard to rigorously prove things about it (and that's interesting in itself—and closely related to the fundamental phenomenon of computational irreducibility). But, for example—like, say, the digits of π—many aspects of it seem random. And, for instance, black and white squares appear to occur with equal frequency—meaning that at the train station the panels let in about 50% of the outside light.

If one looks at sequences of n cells, it seems that all 2^n configurations will occur on average with equal frequency. But not everything is random. And so, for example, if one looks at 3×2 blocks of cells, only 24 of the 32 possible ones ever occur. (Maybe some people waiting for trains will figure out which blocks are missing...)

When we look at the pattern, our visual system particularly picks out the black triangles. And, yes, it seems as if triangles of any size can ultimately occur, albeit with frequency decreasing exponentially with size.

If one looks carefully at the right-hand edge of the rule 30 pattern, one can see that it repeats. However, the repetition period seems to increase exponentially as one goes in from the edge.

At the train station, there are lots of identical panels. But rule 30 is actually an inexhaustible source of new patterns. So what would happen if one just continued the evolution, and rendered it on successive panels? Here's the result. It's a pity about the hint of periodicity on the right-hand edge, and the big triangle on panel 5 (which might be a safety problem at the train station).

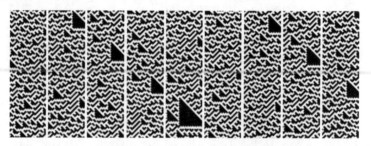

Fifteen more steps in from the edge, there's no hint of that anymore:

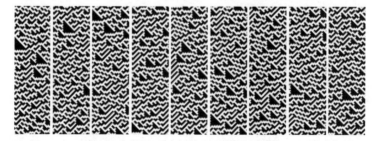

What about other initial conditions? If the initial conditions repeat, then so will the pattern. But otherwise, so far as one can tell, the pattern will look essentially the same as with a single-cell initial condition.

One can try other rules too. Here are a few from the same simplest 256-rule set as rule 30:

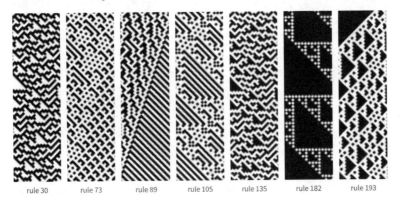

rule 30   rule 73   rule 89   rule 105   rule 135   rule 182   rule 193

Moving deeper from the edge the results look a little different (for aficionados, rule 89 is a transformed version of rule 45, rule 182 of rule 90, and rule 193 of rule 110):

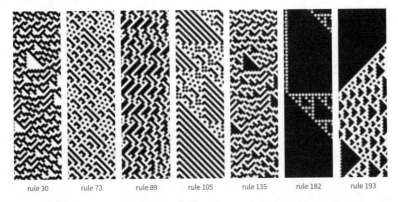

rule 30   rule 73   rule 89   rule 105   rule 135   rule 182   rule 193

And starting from random initial conditions, rather than a single black cell, things again look different:

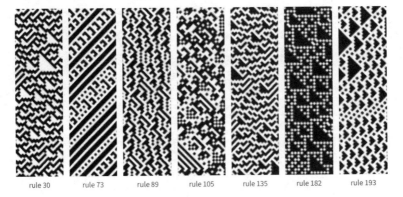

rule 30   rule 73   rule 89   rule 105   rule 135   rule 182   rule 193

And here are a few more rules, started from random initial conditions:

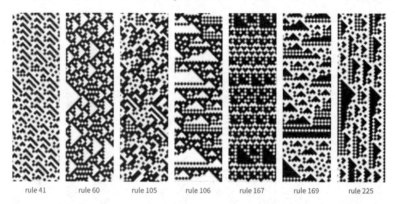

rule 41   rule 60   rule 105   rule 106   rule 167   rule 169   rule 225

It's amazing what's out there in the computational universe of possible programs. There's an infinite range of possible patterns. But it's cool that the Cambridge North train station uses my all-time favorite discovery in the computational universe—rule 30! And it looks great!

The Bigger Picture

There's something curiously timeless about algorithmically generated forms. A dodecahedron from ancient Egypt still looks crisp and modern today. As do periodic tilings—or nested forms—even from centuries ago:

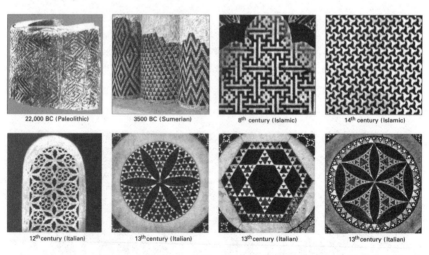

22,000 BC (Paleolithic)   3500 BC (Sumerian)   8th century (Islamic)   14th century (Islamic)

12th century (Italian)   13th century (Italian)   13th century (Italian)   13th century (Italian)

But can one generate richer forms algorithmically? Before I discovered rule 30, I'd always assumed that any form generated from simple rules would always somehow end up being obviously simple. But rule 30 was a big shock to my intuition—and from it I realized that actually in the computational universe of all possible rules, it's actually very easy to get rich and complex behavior, even from simple underlying rules.

And what's more, the patterns that are generated often have remarkable visual interest. Here are a few produced by cellular automata (now with three possible colors for each cell, rather than two):

There's an amazing diversity of forms. And, yes, they're often complicated. But because they're based on simple underlying rules, they always have a certain logic to them: in a sense each of them tells a definite "algorithmic story."

One thing that's notable about forms we see in the computational universe is that they often look a lot like forms we see in nature. And I don't think that's a coincidence. Instead, I think what's going on is that rules in the computational universe capture the essence of laws that govern lots of systems in nature—whether in physics, biology, or wherever. And maybe there's a certain familiarity or comfort associated with forms in the computational universe that comes from their similarity to forms we're used to in nature.

But is what we get from the computational universe art? When we pick out something like rule 30 for a particular purpose, what we're doing is conceptually a bit like photography: we're not creating the underlying forms, but we are selecting the ones we choose to use.

In the computational universe, though, we can be more systematic. Given some aesthetic criterion, we can automatically search through perhaps even millions or billions of possible rules to find optimal ones: in a sense automatically "discovering art" in the computational universe.

We did an experiment on this for music back in 2007: WolframTones. And what's remarkable is that even by sampling fairly small numbers of rules (cellular automata, as it happens), we're able to produce all sorts of interesting short pieces of music—that often seem remarkably "creative" and "inventive."

From a practical point of view, automatic discovery in the computational universe is important because it allows for mass customization. It makes it easy to be "original" (and "creative")—and to find something different every time, or to fit constraints that have never been seen before (say, a pattern in a complicated geometric region).

The Cambridge North train station uses a particular rule from the computational universe to make what amounts to an ornamental pattern. But one can also use rules from the computational universe for other things in architecture. And one can even imagine a building in which everything—from overall massing down to details of moldings—is completely determined by something close to a single rule.

One might assume that such a building would somehow be minimalist and sterile. But the remarkable fact is that this doesn't have to be true—and that instead there are plenty of rich, almost "organic" forms to be "mined" from the computational universe.

Ever since I started writing about one-dimensional cellular automata back in the early 1980s, there's been all sorts of interesting art done with them. Lots of different rules have been used. Sometimes they've been what I called "class 4" rules that have a particularly organic look. But often it's been other rules—and rule 30 has certainly made its share of appearances—whether it's on floors, shirts, tea cosies, kinetic installations, or, recently, mass-customized scarves (with the knitting machine actually running the cellular automaton):

But today we're celebrating a new and different manifestation of rule 30. Formed from permanent aluminum panels, in an ancient university town, a marvellous corner of the computational universe adorns one of the most practical of structures: a small train station. My compliments to the architects. May what they've made give generations of rail travelers a little glimpse of the wonders of the computational universe. And maybe perhaps a few, echoing the last words attributed to the traveler in the movie *2001: A Space Odyssey*, will exclaim, "oh my gosh, it's covered in rule 30s!"

The Personal Analytics of My Life

March 8, 2012

One day I'm sure everyone will routinely collect all sorts of data about themselves. But because I've been interested in data for a very long time, I started doing this long ago. I actually assumed lots of other people were doing it too, but apparently they were not. And so now I have what is probably one of the world's largest collections of personal data.

Every day—in an effort at "self-awareness"—I have automated systems send me a few emails about the day before. But even though I've been accumulating data for years—and always meant to analyze it—I've never actually gotten around to doing it. But with Mathematica and the auto-mated data analysis capabilities we just released in Wolfram|Alpha Pro, I thought now would be a good time to finally try taking a look—and to use myself as an experimental subject for studying what one might call "personal analytics."

Let's start off talking about email. I have a complete archive of all my email going back to 1989—a year after Mathematica was released, and two years after I founded Wolfram Research. Here's a plot with a dot showing the time of each of the third of a million emails I've sent since 1989:

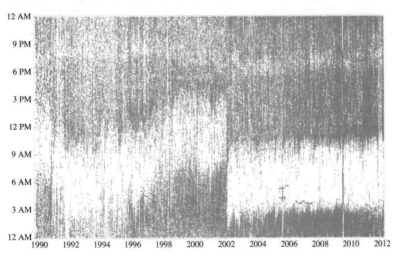

The first thing one sees from this plot is that, yes, I've been busy. And for more than 20 years, I've been sending emails throughout my waking day, albeit with a little dip around dinner time. The big gap each day comes from when I was asleep. And for the last decade, the plot shows I've been pretty consistent, going to sleep around 3am ET, and getting up around 11am (yes, I'm a bit of a night owl). (The stripe in summer 2009 is a trip to Europe.)

But what about the 1990s? Well, that was when I spent a decade as something of a hermit, working very hard on *A New Kind of Science*. And the plot makes it very clear why in the late 1990s when one of my children was asked for an example of "being nocturnal," they gave me. The rather dramatic discontinuity in 2002 is the moment when *A New Kind of Science* was finally finished, and I could start leading a different kind of life.

So what about other features of the plot? Some line up with identifiable events and trends in my life, sometimes reflected in my online scrapbook or timeline. Others at first I don't understand at all—until a quick search of my email archive jogs my memory. It's very convenient that I can always drill down and read a raw email. Because as with essentially any long-timescale data project, there are all kinds of glitches (here like misformatted email headers, unset computer clocks, and untagged automated mailings) that have to be found and systematically corrected for before one has consistent data to analyze. And before, in this case, I can trust that any dots in the middle of the night are actually times I woke up and sent email (which is nowadays very rare).

The previous plot suggests that there's been a progressive increase in my email volume over the years. One can see that more explicitly if one just plots the total number of emails I've sent as a function of time:

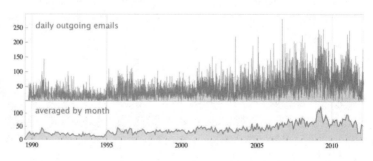

Again, there are some life trends visible. The gradual decrease in the early 1990s reflects me reducing my involvement in day-to-day management of our company to concentrate on basic science. The increase in the 2000s is me jumping back in, and driving more and more company projects. And the peak in early 2009 reflects with the final preparations for the launch of Wolfram|Alpha. (The individual spikes, including the all-time winner August 27, 2006, are mostly weekend or travel days specifically spent "grinding down" email backlogs.)

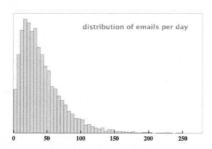

distribution of emails per day

The plots above seem to support the idea that "life's complicated." But if one aggregates the data a bit, it's easy to end up with plots that seem like they could just be the result of some simple physics experiment. Like here's the distribution of the number of emails I've sent per day since 1989:

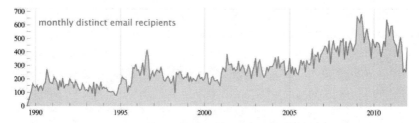

monthly distinct email recipients

What is this distribution? Is there a simple model for it? I don't know. Wolfram|Alpha Pro tells us that the best fit it finds is to a geometric distribution. But it officially rejects that fit. Still, at least the tail seems—as so often—to follow a power law. And perhaps that's telling me something about myself, though I have to say I don't know what.

The vast majority of these recipients are people or mailgroups within our company. And I suspect the overall growth is a reflection of both the increasing number of people at the company, and the increasing number of projects in which I and our company are involved. The peaks are often associated with intense early-stage projects, where I am directly interacting with lots of people, and there isn't yet a well-organized management structure in place. I don't quite understand the recent decrease, considering that the number of projects is at an all-time high. I'm just hoping it reflects better organization and management...

OK, so all of that is about email I've sent. What about email I've received? Here's a plot comparing my incoming and outgoing email:

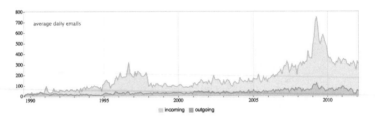

The peaks in 1996 and 2009 are both associated with the later phases of big projects (Mathematica 3 and the launch of Wolfram|Alpha) where I was watching all sorts of details, often using email-based automated systems.

So email is one kind of data I've systematically archived. And there's a huge amount that can be learned from that. Another kind of data that I've been collecting is keystrokes. For many years, I've captured every keystroke I've typed—now more than 100 million of them:

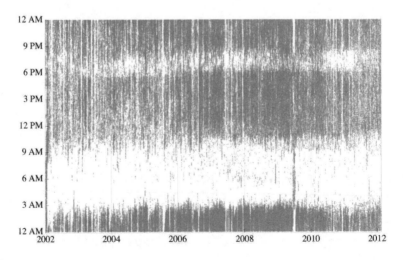

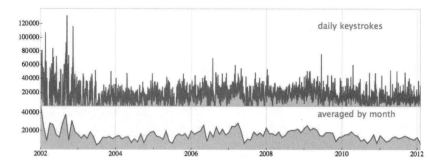

There are all kinds of detailed facts to extract: like that the average fraction of keys I type that are backspaces has consistently been about 7% (I had no idea it was so high!). Or how my habits in using different computers and applications have changed. And looking at the daily totals, I can see spikes of writing activity—typically associated with creating longer documents. But at least at an overall level things like the plots above look similar for keystrokes and email.

What about other measures of activity? My automated systems have been quietly archiving lots of them for years. And for example this shows the times of events that have appeared in my calendar:

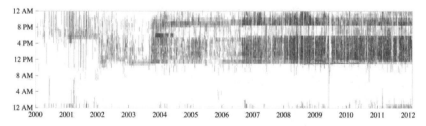

The changes over the years reflect quite directly things going on in my life. Before 2002 I was doing a lot of solitary work, particularly on *A New Kind of Science*, and having only a few scheduled meetings. But then as I initiated more and more new projects at our company, and took a more and more structured approach to managing them, one can see more and more meetings getting filled in—though my "family dinner stripe" remains clearly visible.

Here's a plot of the daily average total number of meetings (and other calendar events) that I've done over the years:

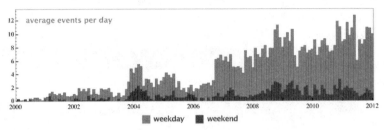

The trend is pretty clear. And it reflects the fact that in the past decade or so I've gradually learned to work better "in public," efficiently figuring things out while interacting with groups of people—which I've discovered makes me much more effective both at using other people's expertise and at delegating things that have to be done.

It often surprises people when I tell them this, but since 1991 I've been a remote CEO, interacting with my company almost exclusively just by email and phone (usually with screensharing). (No, I don't find video-conferencing with the company very useful, and the telepresence robot I got recently has mostly been standing idle.)

So phone calls are another source of data for me. And here's a plot of the times of calls I've made (the light gray vertical bars are missing data):

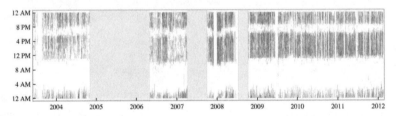

And yes, I spend many hours on the phone each day:

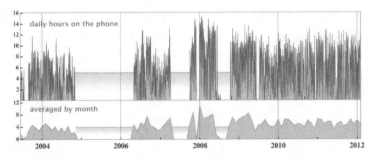

And this shows how the probability to find me on the phone varies during the day:

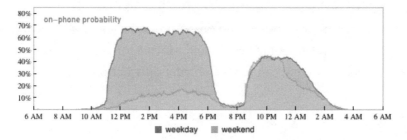

This is averaged over all days for the last several years, and in fact I'm guessing that the "peak weekday probability" would actually be even higher than 70% if the average excluded days when I'm away for one reason or another.

Here's another way to look at the data—this shows the probability for calls to start at a given time:

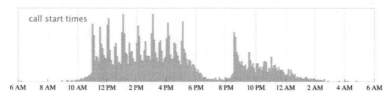

There's a curious pattern of peaks—near hours and half-hours. And of course those occur because many phone calls are scheduled at those times. Which means that if one plots meeting start times and phone call start times one sees a strong correlation:

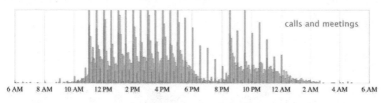

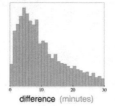

I was curious just how strong this correlation is: in effect just how scheduled all those calls are. And looking at the data I found that at least for my external phone meetings at least half of them do indeed start within two minutes of their appointed times. For internal meetings— which tend to involve more people, and which I normally have scheduled back-to-back—there's a somewhat broader distribution:

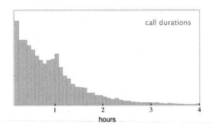

When one looks at the distribution of call durations one sees a kind of "physics-like" background shape, but on top of that there's the "obviously human" peak at the one-hour mark, associated with meetings that are scheduled to be an hour long.

So far everything we've talked about has measured intellectual activity. But I've also got data on physical activity. Like for the past couple of years I've been wearing a little digital pedometer that measures every step I take:

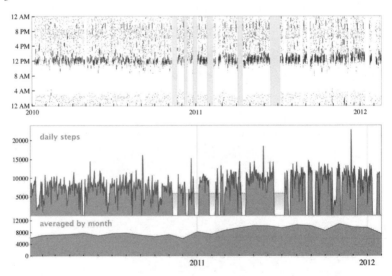

And once again, this shows quite a bit of consistency. I take about the same number of steps every day. And many of them are taken in a block early in my day (typically coinciding with the first couple of meetings I do). There's no mystery to this: years ago I decided I should take some exercise each day, so I set up a computer and phone to use while walking on a treadmill. (Yes, with the correct ergonomic arrangement one can type and use a mouse just fine while walking on a treadmill, at least up to—for me—a speed of about 2.5 mph.)

OK, so let's put all this together. Here are my "average daily rhythms" for the past decade (or in some cases, slightly less):

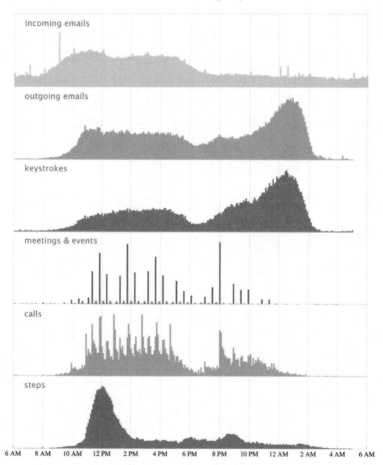

The overall pattern is fairly clear. It's meetings and collaborative work during the day, a dinner-time break, more meetings and collaborative work, and then in the later evening more work on my own. I have to say that looking at all this data I am struck by how shockingly regular many aspects of it are. But in general I am happy to see it. For my consistent experience has been that the more routine I can make the basic practical aspects of my life, the more I am able to be energetic—and spontaneous—about intellectual and other things.

And for me one of the objectives is to have ideas, and hopefully good ones. So can personal analytics help me measure the rate at which that happens?

It might seem very difficult. But as a simple approximation, one can imagine seeing at what rate one starts using new concepts, by looking at when one starts using new words or other linguistic constructs. Inevitably there are tricky issues in identifying genuine new "words" etc. (though for example I have managed to determine that when it comes to ordinary English words, I've typed about 33,000 distinct ones in the past decade). If one restricts to a particular domain, things become a bit easier, and here for example is a plot showing when names of what are now Wolfram Language functions first appeared in my outgoing email:

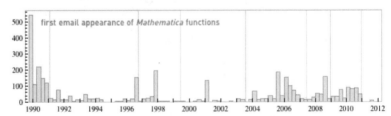

The spike at the beginning is an artifact, reflecting pre-existing functions showing up in my archived email. And the drop at the end reflects the fact that one doesn't yet know future Mathematica names. But it's interesting to see elsewhere in the plot little "bursts of creativity," mostly but not always correlated with important moments in Mathematica history—as well as a general increase in density in recent times.

As a quite different measure of creative progress, here's a plot of when I modified the text of chapters in *A New Kind of Science*:

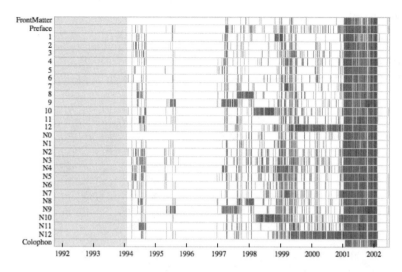

I don't have data readily at hand from the beginning of the project. And in 1995 and 1996 I continued to do research, but stopped editing text, because I was pulled away to finish Mathematica 3 (and the book about it). But otherwise one sees inexorable progress, as I systematically worked out each chapter and each area of the science. One can see the time it took to write each chapter (Chapter 12 on the Principle of Computational Equivalence took longest, at almost two years), and which chapters led to changes in which others. And with enough effort, one could drill down to find out when each discovery was made (it's easier with modern Mathematica automatic history recording). But in the end—over the course of a decade—from all those individual keystrokes and file modifications there gradually emerged the finished *A New Kind of Science*.

It's amazing how much it's possible to figure out by analyzing the various kinds of data I've kept. And in fact, there are many additional kinds of data I haven't even touched on in this chapter. I've also got years of curated medical test data (as well as my not-yet-very-useful complete genome), GPS location tracks, room-by-room motion sensor data, endless corporate records—and much much more.

And as I think about it all, I suppose my greatest regret is that I did not start collecting more data earlier. I have some backups of my computer filesystems going back to 1980. And if I look at the 1.7 million files in my current filesystem, there's a kind of archeology one can do, looking at files that haven't been modified for a long time (the earliest is dated June 29, 1980).

Here's a plot of the latest modification times of all my current files:

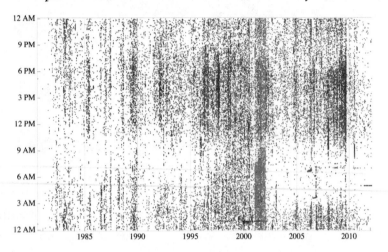

In the early years, there's a mixture of plain text files and C language files. But gradually there's a transition to Mathematica files—with a burst of page layout files from when I was finishing *A New Kind of Science*. And once again the whole plot is a kind of engram—now of more than 30 years of my computing activities.

So what about things that were never on a computer? It so happens that years ago I also started keeping paper documents, pretty much on the theory that it was easier just to keep everything than to worry about what specifically was worth keeping. And now I've got about 230,000 pages of my paper documents scanned, and when possible OCR'ed. And as just one example of the kind of analysis one can do, here's a plot of the frequency with which different four-digit "date-like sequences" occur in all these documents:

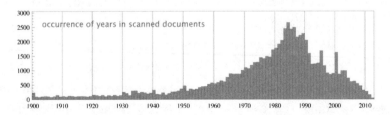

Of course, not all these four-digit sequences refer to dates (especially for example "2000")—but many of them do. And from the plot one can see the rather sudden turnaround in my use of paper in 1984—when I turned the corner to digital storage.

What is the future for personal analytics? There is so much that can be done. Some of it will focus on large-scale trends, some of it on identifying specific events or anomalies, and some of it on extracting "stories" from personal data.

And in time I'm looking forward to being able to ask Wolfram|Alpha all sorts of things about my life and times—and have it immediately generate reports about them. Not only being able to act as an adjunct to my personal memory, but also to be able to do automatic computational history—explaining how and why things happened—and then making projections and predictions.

As personal analytics develops, it's going to give us a whole new dimension to experiencing our lives. At first it all may seem quite nerdy (and certainly as I glance back at this writing there's a risk of that). But it won't be long before it's clear how incredibly useful it all is—and everyone will be doing it, and wondering how they could have ever gotten by before. And wishing they had started sooner, and hadn't "lost" their earlier years.

An updated plot that includes email activity from the past seven years:

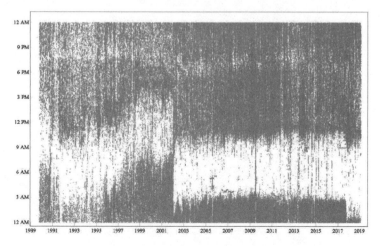

Seeking the Productive Life: Some Details of My Personal Infrastructure

February 21, 2019

The Pursuit of Productivity

I'm a person who's only satisfied if I feel I'm being productive. I like figuring things out. I like making things. And I want to do as much of that as I can. And part of being able to do that is to have the best personal infrastructure I can. Over the years I've been steadily accumulating and implementing "personal infrastructure hacks" for myself. Some of them are, yes, quite nerdy. But they certainly help me be productive. And maybe in time more and more of them will become mainstream, as a few already have.

Now, of course, one giant "productivity hack" that I've been building for the world for a very long time is the whole technology stack around the Wolfram Language. And for me personally, another huge "productivity hack" is my company, which I started more than 32 years ago. Yes, it could (and should) be larger, and have more commercial reach. But as a nicely organized private company with about 800 people it's an awfully efficient machine for turning ideas into real things, and for leveraging what skills I have to greatly amplify my personal productivity.

I could talk about how I lead my life, and how I like to balance doing leadership, doing creative work, interacting with people, and doing things that let me learn. I could talk about how I try to set things up so that what I've already built doesn't keep me so busy I can't start anything new. But instead what I'm going to focus on here is my more practical personal infrastructure: the technology and other things that help me live and work better, feel less busy, and be more productive every day.

At an intellectual level, the key to building this infrastructure is to structure, streamline, and automate everything as much as possible— while recognizing both what's realistic with current technology, and what fits with me personally. In many ways, it's a good, practical exercise in computational thinking, and, yes, it's a good application of some of the tools and ideas that I've spent so long building. Much of it can probably be helpful to lots of other people too; some of it is pretty specific to my personality, my situation, and my patterns of activity.

My Daily Life

To explain my personal infrastructure, I first have to say a bit about my daily life. Something that often surprises people is that for 28 years I've been a remote CEO. I'm about as hands-on a CEO as they come. But I'm only physically "in the office" a few times a year. Mostly I'm just at home, interacting with the company with great intensity—but purely through modern virtual means:

I'm one of those CEOs who actually does a lot of stuff myself, as well as managing other people to do things. Being a remote CEO helps me achieve that, and stay focused. And partly following my example, our company has evolved a very distributed culture, with people working scattered all over the world (it's all about being productive, rather than about "showing up"):

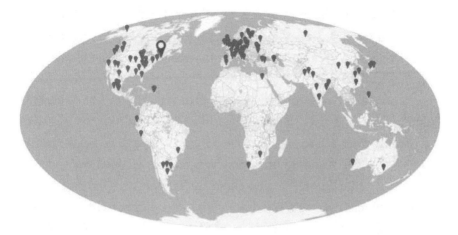

At my desk, though, my basic view of all this is just:

It's always set up the same way. On the right is my main "public display" monitor that I'll be screensharing most of the day with people I'm talking to. On the left is my secondary "private display" monitor that's got my email and messages and other things that aren't directly relevant to the meetings I'm doing.

For the past year or so, I've been livestreaming many of our software design meetings—and there are now 250 hours of archived screensharing, all from that right monitor of mine.

Particularly since I'm at my desk much of each day, I've tried to optimize its ergonomics. The keyboard is at the right height for optimal typing. The monitors are at a height that—especially given my "computer distance" multifocal glasses—forces my head to be in a good position when I look at them, and not hunched over. I still use a "roll-around" mouse (on the left, since I'm left-handed)—because at least according to my latest measurements I'm still faster with that than with any other pointing technology.

At the touch of a button, my desk goes to standing height. But while standing may be better than sitting, I like to at least start my day with something more active, and for more than a decade I've been making sure to walk for a couple of hours every morning. But how can I be productive while I'm walking? Well, nearly 15 years ago (i.e. long before it was popular!) I set up a treadmill with a computer in the room next to my office.

The biomechanics weren't too hard to work out. I found out that by putting a gel strip at the correct pivot point under my wrists (and putting the mouse on a platform) I can comfortably type while I'm walking. I typically use a 5% incline and go at 2 mph—and I'm at least fit enough that I don't think anyone can tell I'm walking while I'm talking in a meeting. (And, yes, I try to get potentially frustrating meetings scheduled during my walking time, so if I do in fact get frustrated I can just "walk it off" by making the treadmill go a little faster.)

For many years I've kept all kinds of personal analytics data on myself, and for the past couple of years this has included continuous heart-rate data. Early last summer I noticed that for a couple of weeks my resting heart rate had noticeably gone down. At first I thought it was just because I happened to be systematically doing something I liked then. But later in the summer, it happened again. And then I realized: those were times when I wasn't walking inside on a treadmill; instead (for different reasons) I was walking outside.

For many years my wife had been extolling the virtues of spending time outside. But it had never really seemed practical for me. Yes, I could talk on the phone (or, in rare cases, actually talk to someone I was walking with). Or I could be walking with a tablet, perhaps watching someone else screensharing—as I did, rather unstylishly, for a week late last summer during my version of a vacation:

I'd actually been thinking about walking and working for a long time. Twenty years ago I imagined doing it with an augmented reality display and a one-handed (chorded) keyboard. But the technology didn't arrive, and I wasn't even sure the ergonomics would work out (would it make me motion sick, for example?).

But then, last spring, I was at a fancy tech event, and I happened to be just out of the frame of a photo op that involved Jeff Bezos walking with a robotic dog. I wasn't personally so excited about the robotic dog. But what really interested me was the person walking out of the frame on the other side, intently controlling the dog—using a laptop that he had strapped on in front of him as if he were selling popcorn.

Could one actually work like this, typing and everything? After my "heart-rate discovery" I decided I had to try it. I thought I'd have to build something myself, but actually one can just buy "walking desks", and so I did. And after minor modifications, I discovered that I could walk and type perfectly well with it, even for a couple of hours. I was embarrassed I hadn't figured out such a simple solution 20 years ago. But starting last fall—whenever the weather's been good—I've tried to spend a couple of hours of each day walking outside like this:

And even when I'm intently concentrating on my computer, it's somehow nice to be outside—and, yes, it seems to have made my resting heart rate go down. And I seem to have enough peripheral vision—or perhaps I've just been walking in "simple enough" environments—that I haven't tripped even when I'm not consciously paying attention. No doubt it helps that I haven't mostly been walking in public places, so there aren't other people around. Of course, that also means that I haven't had the opportunity to get the kind of curious stares I did in 1987 when I first walked down a city street talking on a shoe-sized cellphone....

My Desk Environment

I've had the same big wooden desk for 25 years. And needless to say, I had it constructed with some special features. One of my theories of personal organization is that any flat surface represents a potential "stagnation point" that will tend to accumulate piles of stuff—and the best way to avoid such piles is just to avoid having permanent flat surfaces. But one inevitably needs some flat surface, if only just to sign things (it's not all digital yet), or to eat a snack. So my solution is to have pullouts. If one needs them, pull them out. But one can't leave them pulled out, so nothing can accumulate on them:

These days I don't deal with paper much. But whenever something does come across my desk, I like to file it. So behind my desk I have an array of drawers—with the little hack that there's a slot at the top of each drawer that allows me to immediately slide things into the drawer, without opening it:

I used to fill up a banker's box with filed papers every couple of months; now it seems to take a couple of years. And perhaps as a sign of how paperless I've become, I have a printer under my desk that I use so rarely that I now seem to go through a ream of paper only every year or so.

There are also other things that have changed over the years. I always want my main computer to be as powerful as possible. And for years that meant that it had to have a big fan to dissipate heat. But since I really like my office to be perfectly quiet (it adds a certain calmness that helps my concentration), I had to put the CPU part of my computer in a different room. And to achieve this, I had a conduit in the floor, through which I had to run often-finicky long-distance video cables. Well, now, finally, I have a powerful computer that doesn't need a big fan—and so I just keep it behind my desk. (I actually also have three other not-so-quiet computers that I keep in the same room as the treadmill, so that when I'm on the treadmill I can experience all three main modern computing environments, choosing between them with a KVM switch.)

When I mention to people that I'm a remote CEO, they often say, "You must do lots of videoconferencing." Well, actually, I do basically no videoconferencing. Screensharing is great, and critical. But typically I find video distracting. Often I'll do a meeting where I have lots of people in case we need to get their input. But for most of the meeting I don't need all of them to be paying attention (and I'm happy if they're getting other work done). But if video is on, seeing people who are not paying attention just seems to viscerally kill the mood of almost any meeting.

Given that I don't have video, audio is very important, and I'm quite a stickler for audio quality in meetings. No speakerphones. No bad cell-phone connections. I myself remain quite old school. I wear a headset (with padding added to compensate for my lack of top-of-head hair) with a standard boom microphone. And—partly out of caution about having a radio transmitter next to my head all day—my headset is wired, albeit with a long wire that lets me roam around my office.

Even though I don't use "talking head" video for meetings, I do have a document camera next to my computer. One time I'll use this is when we're talking about phones or tablets. Yes, I could connect their video directly into my computer. But if we're discussing user experience on a phone it's often helpful to be able to actually see my finger physically touching the phone.

The document camera also comes in handy when I want to show pages from a physical book, or artifacts of various kinds. When I want to draw something simple I'll use the annotation capabilities of our screen-sharing system. But when I'm trying to draw something more elabo-rate I'll usually do the retro thing of putting a piece of paper under the document camera, then just using a pen. I like the fact that the image from the document camera comes up in a window on my screen, that I can resize however I want. (I periodically try using drawing tablets but I don't like the way they treat my whole screen as a canvas, rather than operating in a window that I can move around.)

On the Move

In some ways I lead a simple life, mostly at my desk. But there are plenty of times when I'm away from my desk—like when I'm someplace else in my house, or walking outside. And in those cases I'll normally take a 13" laptop to use. When I go further afield, it gets a bit more complicated.

If I'm going to do serious work, or give a talk, I'll take the 13" laptop. But I never like to be computerless, and the 13" laptop is a heavy thing to lug around. So instead I also have a tiny 2-lb laptop, which I put in a little bag (needless to say, both the bag and the computer are adorned with our Spikey logo):

And for at least the past couple of years—unless I'm bringing the bigger computer, usually in a backpack—I have taken to "wearing" my little computer wherever I go. I originally wanted a bag where the computer would fit completely inside, but the nicest bag I could find had the computer sticking out a bit. To my surprise, though, this has worked well. And it's certainly amusing when I'm talking to someone and quickly "draw" my computer, and they look confused, and ask, "Where did that come from?"

I always have my phone in my pocket, and if I have just a few moments that's what I'll pull out. It works fine if I'm checking mail, and deleting or forwarding a few messages. If I actually want to write anything serious, though, out will come my little computer, with its full keyboard. Of course, if I'm standing up it's pretty impractical to try to balance the computer on one hand and type with the other. And sometimes if I know I'm going to be standing for a while, I'll bring a tablet with me. But other times, I'll just be stuck with my phone. And if I run out of current things I can usefully do (or I don't have an internet connection) I'll typically start looking at the "things to read" folder that I maintain synched on all my devices.

Back in 2007 I invented WolframTones because I wanted to have a unique ringtone for my phone. But while WolframTones has been successful as an example of algorithmic music composition, the only trace of it on my phone is the image of WolframTones compositions that I use as my home screen:

How do I take notes when I'm "out and about"? I've tried various technological solutions, but in the end none have proved both practical and universally socially acceptable. So I've kept doing the same thing for 40 years: in my pocket I have a pen, together with a piece of paper folded three times (so it's about the size of a credit card). It's very low-tech, but it works. And when I come back from being out I always take a few moments to transcribe what I wrote down, send out emails, or whatever.

I have little "tech survival kits" that I bring with me. The centerpiece is a tiny charger, that charges both my computer (through USB-C) and my phone. I bring various connectors, notably so I can connect to things like projectors. I also bring a very light 2- to 3-prong power adaptor, so I don't find my charger falling out of overused power outlets.

When I'm going on "more serious expeditions" I'll add some things to the kit: there's a "charging brick" (unfortunately now in short supply) that'll keep my computer going for many hours. For events like trade shows, I'll bring a tiny camera that takes pictures every 30 seconds, so 1 can remember what 1 saw. And if I'm really going out into the wilds, I'll bring a satphone as well. (Of course, 1 always have other stuff too, like a very thin and floppy hat, a light neoprene bag-within-a-bag, glasses wipes, hand sanitizer, mosquito wipes, business cards, pieces of chocolate, etc.)

In my efforts to keep organized on trips, I'll typically pack several plastic envelopes:

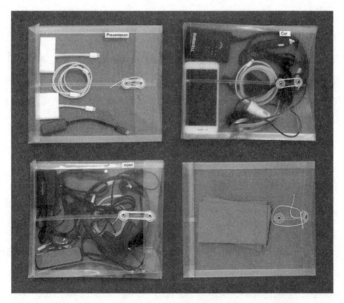

In "Presentation" there'll be the adaptors (VGA, HDMI, ...) 1 need to connect to projectors. Sometimes there'll be a wired Ethernet adaptor. (For very low-key presentations, I'll also sometimes bring a tiny projector too.) In "Car" there'll be a second cellphone that can be used as a GPS, with a magnetic back and a tiny thing for attaching to the air vent in a car. There'll be a monaural headset, a phone charger, and sometimes a tiny inverter for my computer. If I'm bringing the satphone, there'll also be a car kit for it, with an antenna that magnets to the roof of the car, so it can "see" the satellites. In "Hotel" there'll be a binaural headset, a second computer charger, and a disk with an encrypted backup of my computer, in case I lose my computer and have to buy and configure a new machine. The fourth plastic envelope is used to store things I get on the trip, and it contains little envelopes—approximately one for each day of my trip—in which I put business cards.

Years ago, I always used to bring a little white-noise fan with me, to mask background noise, particularly at night. But at some point I realized that I didn't need a physical fan, and instead I just have an app that simulates it (I used to use pink noise, but now I just use "air conditioner sound"). It's often something of a challenge to predict just how loud the outside noise one's going to encounter (say, the next morning) will be, and so how loud one should set the masking sound. And, actually, as I write this, I realize I should use modern audio processing in the Wolfram Language to just listen to external sounds, and adjust the masking sound to cover them.

Another thing I need when I travel is a clock. And nowadays it's just a piece of Wolfram Language code running on my computer. But because it's software, it can have a few extra features. I always leave my computer on my home timezone, so the "clock" has a slider to specify local time (yes, if I'm ever in a half-hour timezone again I'll have to tweak the code). It also has a button **Start sleep timer**. When I press it, it starts a count-up timer, which lets me see how long I've been asleep, whatever my biological clock may say. (**Start sleep timer** also sends an email which gives my assistant an idea of whether or not I'll make it to that early-next-morning meeting. The top right-hand "mouse corner" is a hack for preventing the computer from going to sleep.)

Whenever it's practical, I like to drive myself places. It was a different story before cellphones. But nowadays if I'm driving I'm productively making a phone call. I'll have meetings that don't require me to look at anything scheduled for my "drive times" (and, yes, it's nice to have standard conference call numbers programmed in my phone, so I can voice-dial them). And I maintain a "call-while-driving" list of calls that I can do while driving, particularly if I'm in an unusual-for-me timezone.

I've always had the problem that if I try to work on a computer while I'm being driven by someone else, I get car sick. I thought I had tried everything. Big cars. Little cars. Hard suspension. Soft suspension. Front seat. Back seat. Nothing worked. But a couple of years ago, quite by chance, I tried listening to music with big noise-canceling head-phones—and I didn't get car sick. But what if when I'm being driven I want to be on the phone while I'm using my computer? Well, at the 2018 Consumer Electronics Show, despite my son's admonition that "just because you can't tell what they're selling at a booth doesn't mean it's interesting," I stopped at a booth and got these strange objects, which, despite looking a bit odd, do seem to prevent car sickness for me, at least much of the time:

Giving Talks

I give quite a lot of talks—to a very wide range of audiences. I particularly like giving talks about subjects I haven't talked about before. I give talks to the fanciest business, tech, and science groups. I give talks to schoolkids. I enjoy interacting with audiences (Q&A is always my favorite part), and I enjoy being spontaneous. And I essentially always end up doing livecoding.

When I was young I traveled quite a bit. I did have portable computers even back in the 1980s (my first was an Osborne 1 in 1981), though mostly in those days my only way to stay computer-productive was to have workstation computers shipped to my destinations. Then in the early 1990s, I decided I wasn't going to travel anymore (not least because I was working so intensely on *A New Kind of Science*). So for a while I basically didn't give any talks. But then technology advanced. And it started being realistic to give talks through videoconferencing.

I went through several generations of technology, but a number of years ago I built out a videoconferencing setup in my basement. The "set" can be reconfigured in various ways (podium, desk, etc.) But basically I have a back-projection screen on which I can see the remote audience. The camera is in front of the screen, positioned so I'm looking straight at it. If I'm using notes or a script (which, realistically, is rare) I have a homemade teleprompter consisting of a half-silvered mirror and a laptop that I can look at the camera through.

While it's technically feasible for me to be looking straight at the camera when I'm livecoding, this makes it look to the audience as if I'm staring off into space, which seems weird. It's better to look slightly down when I'm obviously looking at a screen. And in fact with some setups it's good for the audience to see the top of a computer right at the bottom of the screen, to "explain" what I'm looking at.

Videoconferenced talks work quite well in many settings (and, for some extra fun, I've sometimes used a telepresence robot). But in recent years (partly as a result of my children wanting to do it with me) I've decided that traveling is OK—and I've been all over the place:

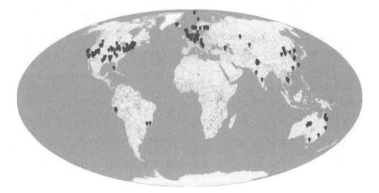

I'll usually be giving talks—often several per day. And I've gradually developed an elaborate checklist of what's needed to have them work. A podium that's at the right height and horizontal enough to let me type easily on my computer (and preferably not so massive that I'm hidden from the audience). An attachable microphone that leaves my hands free to type. A network connection that lets me reach our servers. And, of course, to let the audience actually see things, a computer projector.

I remember the very first computer projector I used, in 1980. It was a Hughes "liquid crystal light valve", and once I got it connected to a CRT terminal, it worked beautifully. In the years since then I've used computer projectors all over the world, both in the fanciest audiovisual situations, and in outlying places with ancient equipment and poor infrastructure. And it's amazing how random it is. In places where one can't imagine the projector is going to work, it'll be just fine. And in places where one can't imagine it won't work, it'll fail horribly.

Some years ago I was giving a talk at TED—with some of the fanciest audiovisual equipment I'd ever seen. And that was one of the places where things failed horribly. Fortunately we did a test the day before. But it took a solid three hours to get the top-of-the-line computer projector to successfully project my computer's screen.

And as a result of that very experience I decided I'd better actually understand how computers talk to projectors. It's a complicated business, that involves having the computer and the projector negotiate to find a resolution, aspect ratio, frame rate, etc. that will work for both of them. Underneath, there are things called EDID strings that are exchanged, and these are what typically get tangled up. Computer operating systems have gotten much better about handling this in recent years, but for high-profile, high-production-value events, I have a little box that spoofs EDID strings to force my computer to send a specific signal, regardless of what the projector seems to be asking it for.

Some of the talks I give are completely spontaneous. But often I'll have notes—and occasionally even a script. And I'll always write these in a Wolfram Notebook. I then have code that "paginates" them, basically replicating "paragraphs" at the end of each page, so I have freedom in when I "turn the page." In past years I used to transfer these notes to an iPad that I'd set up to "turn the page" whenever I touched its screen. But in recent years I've actually just synched files, and used my little computer for my notes—which has the advantage that I can edit them right up to the moment I start giving the talk.

In addition to notes, I'll sometimes also have material that I want to immediately bring into the talk. Now that we have our new Presenter Tools system, I may start creating more slide-show-like material. But that's not how I've traditionally worked. Instead, I'll typically just have a specific piece of Wolfram Language code I want to input, without having to take the time to explicitly type it. Or perhaps I'll want to pick an image from a "slide farm" that I want to immediately put on the screen, say in response to a question. (There's a lot of trickiness about projector resolutions in, for example, slides of cellular automata, because unless they're "pixel perfect" they'll alias—and it's not good enough just to scale them like typical slide software would.)

So how do I deal with bringing in this material? Well, I have a second display connected to my computer—whose image isn't projected. (And, yes, this can contribute to horrible tangling of EDID strings.) Then on that second display I can have things to click or copy. (I have a Wolfram Language function that will take a notebook of inputs and URLs, and make me a palette that I can click to type inputs, open webpages, etc.)

In the past we used to have a little second monitor to attach to my laptop—essentially a disembodied laptop screen. But it took all sorts of kludges to get both it and the projector connected to my laptop (sometimes one would be USB, one would be HDMI, etc.) But now we can just use an iPad—and it's all pure software (though the interaction with projectors can still be finicky):

For a while, just to be stylish, I was using a computer with a Spikey carved out of its case, and backlit. But the little rhombuses in it were a bit fragile, so nowadays I mostly just use "Spikey skins" on my computers:

My Filesystem

The three main applications I use all day are Wolfram Desktop, a web browser, and email. My main way of working is to create (or edit) Wolfram Notebooks. Here are a few notebooks I worked on today:

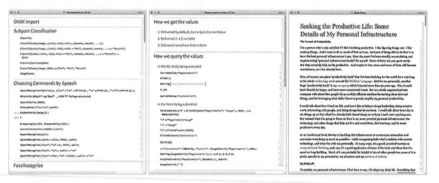

On a good day I'll type at least 25,000 characters into Wolfram Notebooks (and, yes, I record all my keystrokes). I always organize my notebooks into sections and subsections and so on (which, very conveniently, automatically exist in hierarchical cells). Sometimes I'll write mostly text in a notebook. Sometimes I'll screen capture something from elsewhere and paste it in, as a way to keep notes. Depending on what I'm doing, I'll also actually do computations in a notebook, entering Wolfram Language input, getting results, etc.

Over the years, I've accumulated over a hundred thousand notebooks, representing product designs, plans, research, writings, and, basically, everything I do. All these notebooks are ultimately stored in my filesystem (yes, I sync with the cloud, use cloud files, and file servers, etc.) And I take pains to keep my filesystem organized—with the result I can typically find any notebook I'm looking for just by navigating my filesystem, faster than I could formulate a search for it.

I believe I first thought seriously about how to organize my files back in 1978 (which was also when I started using the Unix operating system). And over the past 40 years I've basically gone through five generations of filesystem organization, with each generation basically being a reflection of how I'm organizing my work at that stage in my life.

For example, during the period from 1991 to 2002 when I was writing my big book *A New Kind of Science*, a substantial part of my filesystem was organized simply according to sections of the book:

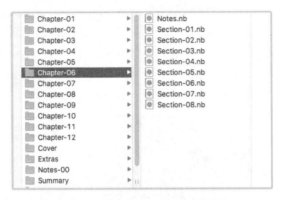

And it's very satisfying that today I can go immediately from, say, an image in the online version of the book, to the notebook that created it (and the stability of the Wolfram Language means that I can immediately run the code in the notebook again—though sometimes it can now be written in a more streamlined way).

The sections of the book are basically laid out in the NewScience/Book/ Layout/ folder of my "third-generation" filesystem. Another part of that filesystem is NewScience/BookResearch/Topics. And in this folder are about 60 subfolders named for broad topics that I studied while working on the book. Within each of these folders are then further subfolders for particular projects I did while studying those topics—which often then turned into particular sections or notes in the book.

Some of my thinking about computer filesystems derives from my experience in the 1970s and 1980s with physical filesystems. Back when I was a teenager doing physics I voraciously made photocopies of papers. And at first I thought the best way to file these papers would be in lots of different categories, with each category stored in a different physical file folder. I thought hard about the categories, often feeling quite pleased with the cleverness of associating a particular paper with a particular category. And I had the principle that if too many papers accumulated in one category, I should break it up into new categories.

All this at first seemed like a good idea. But fairly quickly I realized it wasn't. Because too often when I wanted to find a particular paper I couldn't figure out just what cleverness had caused me to associate it with what category. And the result was that I completely changed my approach. Instead of insisting on narrow categories, I allowed broad, general categories—with the result that I could easily have 50 or more papers filed in a single category (often ending up with multiple well-stuffed physical file folders for a given category):

And, yes, that meant that I would sometimes have to leaf through 50 papers or more to find one I wanted. But realistically this wouldn't take more than a few minutes. And even if it happened several times a day it was still a huge win, because it meant that I could actually successfully find the things I wanted.

I have pretty much the same principle about some parts of my computer filesystem today. For example, when I'm collecting research about some topic, I'll just toss all of it into a folder named for that topic. Sometimes I'll even do this for years. Then when I'm ready to work on that topic, I'll go through the folder and pick out what I want.

These days my filesystem is broken into an active part (that I continuously sync onto all my computers), and a more archival part, that I keep on a central fileserver (and that, for example, contains my older-generation filesystems).

There are only a few top-level folders in my active filesystem. One is called Events. Its subfolders are years. And within each year I'll have a folder for each of the outside events I go to in that year. In that folder I'll store material about the event, notebooks I used for talks there, notes I made at the event, etc. Since in a given year I won't go to more than, maybe, 50 events, it's easy to scan through the Events folder for a given year, and find the folder for a particular event.

Another top-level folder is called Designs. It contains all my notes about my design work on the Wolfram Language and other things we're building. Right now there are about 150 folders about different active areas of design. But there's also a folder called ARCHIVES, which contains folders about earlier areas that are no longer active.

And in fact this is a general principle in the project-oriented parts of my filesystem. Every folder has a subfolder called ARCHIVES. I try to make sure that the files (or subfolders) in the main folder are always somehow active or pending; anything that's finished with I put in ARCHIVES. (I put the name in capitals so it stands out in directory listings.)

For most projects I'll never look at anything in ARCHIVES again. But of course it's easy to do so if I want to. And the fact that it's easy is important, because it means I don't have nagging concerns about saying "this is finished with; let's put it in ARCHIVES," even if I think there's some chance it might become active again.

As it happens, this approach is somewhat inspired by something I saw done with physical documents. When I was consulting at Bell Labs in the early 1980s I saw that a friend of mine had two garbage cans in his office. When I asked him why, he explained that one was for genuine garbage and the other was a buffer into which he would throw documents that he thought he'd probably never want again. He'd let the buffer garbage can fill up, and once it was full, he'd throw away the lower documents in it, since from the fact that he hadn't fished them out, he figured he'd probably never miss them if they were thrown away permanently.

Needless to say, I don't follow exactly this approach, and in fact I keep everything, digital or paper. But the point is that the ARCHIVES mechanism gives me a way to easily keep material while still making it easy to see everything that's active.

I have a bunch of other conventions too. When I'm doing designs, I'll typically keep my notes in files with names like Notes-01.nb or SWNotes-01.nb. It's like my principle of not having too many file categories: I don't tend to try to categorize different parts of the design. I just sequentially number my files, because typically it'll be the most recent—or most recent few—that are the most relevant when I continue with a particular design. And if the files are just numbered sequentially, it's easy to find them; one's not trying to remember what name one happened to give to some particular direction or idea.

A long time ago I started always naming my sequential files file-01, file-02, etc. That way pretty much any sorting scheme will sort the files in sequence. And, yes, I do often get to file-10, etc. But in all these years I have yet to get even close to file-99.

Knowing Where to Put Everything

When I'm specifically working on a particular project, I'll usually just be using files in the folder associated with that project. But on a good day, I'll have lots of ideas about lots of different projects. And I also get hundreds of emails every day, relevant to all sorts of different projects. But often it'll be months or years before I'm finally ready to seriously concentrate on one of these other projects. So what I want to do is to store the material I accumulate in such a way that even long in the future I can readily find it.

For me, there are typically two dimensions to where something should be stored. The first is (not surprisingly) the content of what it's about. But the second is the type of project in which I might use it. Is it going to be relevant to some feature of some product? Is it going to be raw material for some piece I write? Is it a seed for a student project, say at our annual Summer School? And so on.

For some types of projects, the material I'm storing typically consists of a whole file, or several files. For others, I just need to store an idea which can be summarized in a few words or paragraphs. So, for example, the seed for a student project is typically just an idea, that I can describe with a title, and perhaps a few lines of explanation. And in any given year I just keep adding such project ideas to a single notebook—which, for example, I'll look at—and summarize—right before our annual summer programs.

For pieces like this that I'm potentially going to write, it's a little different. At any given time, there are perhaps 50 pieces that I'm considering at some point writing. And what I do is to create a folder for each of them. Each will typically have files with names like Notes-01.nb, into which I accumulate specific ideas. But then the folder will also contain complete files, or groups of files, that I accumulate about the topic of the piece. (Sometimes I'll organize these into subfolders, with names like Explorations and Materials.)

In my filesystem, I have folders for different types of projects: Writings, Designs, StudentProjects, etc. I find it important to have only a modest number of such folders (even with my fairly complex life, not much more than a dozen). When something comes in—say from a piece of email, or from a conversation, or from something I see on the web, or just from an idea I have—I need to be able to quickly figure out what type of project (if any) it might be relevant to.

At some level it's as simple as "what file should I put it into?" But the key point is to have a pre-existing structure that makes it quick to decide that—and then to have this structure be one in which I can readily find things even far into the future.

There are plenty of tricky issues. Particularly if years go by, the way one names or thinks about a topic may change. And sometimes that means at some point I'll just rename a folder or some such. But the crucial thing as far as I'm concerned is that at any given time the total number of folders into which I'm actively putting things is small enough that I can basically remember all of them. I might have a dozen folders for different types of projects. Then some of these will need subfolders for specific projects about specific topics. But I try to limit the total number of "active accumulation folders" to at most a few hundred.

Some of those "accumulation folders" I've had for a decade or more. A few will come into existence and be gone within a few months. But most will last at most a few years—basically the time between when I conceptualize a project, and when the project is, for practical purposes, finished.

It's not perfect, but I end up maintaining two hierarchies of folders. The first, and most important, is in my filesystem. But the second is in my email. There are two basic reasons I maintain material in email folders. The first is immediate convenience. Some piece of mail comes in and I think "that's relevant to such-and-such a project that I'm planning to do"—and I want to store it in an appropriate place. Well, if that place is a mail folder, all I have to do is move the mail with one mouse motion (or maybe with one press of a Touch Bar button). I don't have to, for example, find a file or filesystem folder to put it into.

There's also another reason it's good to leave mail as mail: threading. In the Wolfram Language we've now got capabilities both for importing mailboxes, and for connecting to live mail servers. And one of the things one quickly sees is how complicated the graphs (actually, hypergraphs) of email conversations can be. Mail clients certainly aren't perfect as a way to view these conversations, but it's a lot better to use one than, say, to have a collection of separate files.

When projects are fairly well defined, but aren't yet very active, I tend to use filesystem folders rather than email folders. Typically what will be coming in about these projects are fairly isolated (and non-threaded) pieces of mail. And I find it best either just to drag those pieces of mail into appropriate project folders, or to copy out their contents and add them to notebooks.

When a project is very active, there may be lots of mail coming in about it, and it's important to preserve the threading structure. And when a project isn't yet so well defined, I just want to throw everything about it into a single "bucket," and not have to think about organizing it into subfolders, notebooks, etc.

If I look at my mail folders, I see many that parallel folders in my filesystem. But I see some that do not, particularly related to longer-term project concepts. And I have many such folders that have been there for well over a decade (my current overall mail folder organization is about 15 years old). Sometimes their names aren't perfect. But there are few enough folders, and I've seen them for long enough, that I have a sense of what I'm filing in them, even though their names don't quite capture it.

It's always very satisfying when I'm ready to work on a project, and I open the mail folder for it, and start going through messages, often from long ago. Just in the past few weeks, as we wrap up a major new version of the Wolfram Language, I'm starting to look ahead, and I've been going through folders with messages from 2005, and so on. When I saved those messages, I didn't yet have a definite framework for the project they're about. But now I do. So when I go through the messages I can quickly put them into the appropriate active notebooks and so on. Then I delete the messages from the mail folder, and eventually, once it is empty, delete the whole mail folder. (Unlike with files, I don't find it useful to have an ARCHIVES folder for mail; the mail is just too voluminous and not organized enough, so to find any particular item I'll probably end up having to search for it anyway, and of course I certainly have all of my mail stored.)

OK, so I have my filesystem, and I have mail. At our company we also have an extensive project management system, as well as all sorts of databases, request trackers, source control systems, etc. Mostly the nature of my current work does not cause me to interact directly with these, and I don't explicitly store my own personal output in them. At different times, and with different projects, I have done so. But right now my interaction with these systems is basically only as a viewer, not an author.

Beyond these systems, there are lots of things that I interact with basically through webpages. These might be public sites like wolframalpha. com or wolfram.com. They might be internal sites at our company. And they might be preliminary (say, "test" or "devel") versions of what will in the future be public websites or web-based services. I have a personal homepage that gives me convenient access to all these things:

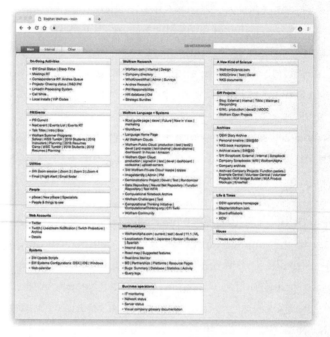

The source for the homepage is (needless to say) a Wolfram Notebook. I can edit this notebook in my filesystem, then press a button to deploy a version to the Wolfram Cloud. I've got an extension in my web browser so that every time I create a new browser window or tab, the initial content will be my personal homepage.

And when I'm going to start doing something, there are just a few places I go. One is this web homepage, which I access many hundreds of times every day. Another is my email and its folders. Another is my desktop filesystem. And basically the only other one of any significance is my calendar system.

From time to time, I'll see other people's computers, and their desktops will be full of files. My desktop is completely empty, and plain white (convenient for whole-screen screensharing and livestreaming). I'd be mortified if there were any files to be seen on my desktop. I'd consider it a sign of defeat in my effort to keep what I'm doing organized. The same can be said of generic folders like Documents and Downloads. Yes, in some situations applications etc. will put files there. But I consider these directories to be throwaways. Nothing in them do I intend to be part of my long-term organizational structure. And they're not synched to the cloud, or across my different computers.

Whatever the organization of my files may be, one feature of them is that I keep them a long time. In fact, my oldest file dates are from 1980. Back then, there was something a bit like the cloud, except it was called timesharing. I've actually lost some of the files that I had on timesharing systems. But the ones I had on on-premise computers are still with me (though, to be fair, some had to be retrieved from 9-track backup tapes).

And today, I make a point of having all my files (and all my email) actively stored on-premise. And, yes, that means I have this in my basement:

The initial storage is on a standard RAID disk array. This is backed up to computers at my company headquarters (about 1000 miles away), where standard tape backups are done. (In all these years, I've only ever had to retrieve from a backup tape once.) I also sync my more active files to the cloud, and to all my various computers.

All the Little Conveniences

My major two personal forms of output are mail messages and Wolfram Notebooks. And over the 30 years since we first introduced notebooks we've optimized our notebook system to the point where I can just press a key to create a default new notebook, and then I'm immediately off and running writing what automatically becomes a good-looking structured document. (And, by the way, it's very nice to see that we've successfully maintained compatibility for 30 years: notebooks I created back in 1988 still just work.)

Sometimes, however, I'm making a notebook that's not so much for human consumption as for input to some automated process. And for this, I use a whole variety of specially set up notebooks. For example, if I want to create an entry in our new Wolfram Function Repository, I just go to the menu item (available in any Version 12 system) File > New > Repository Item > Function Repository Item:

This effectively "prompts" me for items and sections to add. When I'm done, I can press Submit to Repository to send the notebook off to our central queue for repository item reviews (and, just because I'm the CEO doesn't mean I get out of the review process—or want to).

I actually create a fair amount of content that's structured for further processing. A big category is Wolfram Language documentation. And for authoring this we have an internal system we call DocuTools, that's all based on a giant palette developed over many years, that I often say reminds one of an airplane cockpit in its complexity:

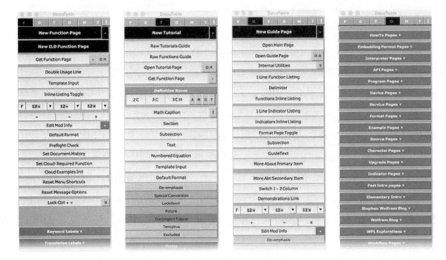

The idea of DocuTools is to make it as ergonomic as possible to author documentation. It has more than 50 subpalettes (a few shown above), and altogether no less than 1016 buttons. If I want to start a new page for a Wolfram Language function I just press New Function Page, and up pops:

A very important part of this page is the stripe at the top that says "Future." This means that even though the page will be stored in our source control system, it's not ready yet: it's just something we're considering for the future. And the system that builds our official documentation will ignore the page.

Usually we (which quite often actually means me) will write documentation for a function before the function is implemented. And we'll include all sorts of details about features the function should have. But when the function is actually first implemented, some of those features may not be ready yet. And to deal with this we (as we call it) "futurize" parts of the documentation, giving it a very visible pink background (light gray highlighting). It's still there in the source control system, and we see it every time we look at the source for the documentation page. But it's not included when the page for documentation that people will see is built.

DocuTools is of course implemented in the Wolfram Language, making extensive use of the symbolic structure of Wolfram Notebooks.

And over the years it's grown to handle many things that aren't strictly documentation; in fact, for me it's become the main hub for the creation of almost all notebook-based content.

There's a button, for example, for Stephen Wolfram Blog. Press it and one gets a standard notebook ready to write into. But in DocuTools there's a whole array of buttons that allow one to insert suggestions and edits. And when I've written a blog what will come back is typically something like this:

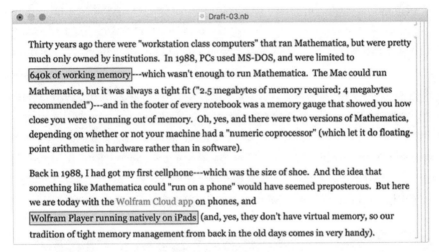

Some boxes are "you really need to fix this"; others are "here's a comment." Click one and up comes a little form:

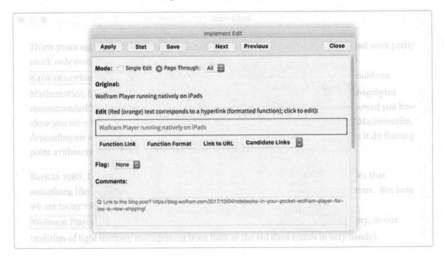

Of course, there are plenty of change-tracking and redlining systems out there in the world. But with the Wolfram Language it becomes easy to create a custom one that's optimized for my needs, so that's what I've had done. Before I had this, it used to take many hours to go through edit suggestions (I remember a horrifying 17-hour plane ride where I spent almost the whole time going through suggestions for a single post). But now—because it's all optimized for me—I can zip through perhaps 10 times faster.

Very often tools that are custom built for me end up being adapted so everyone else can use them too. An example is a system for authoring courses and creating videos. I wanted to be able to do this as a "one-man band"—a bit like how I do livestreaming. My idea was to create a script that contains both words to say and code to input, then to make the video by screen recording in real time while I went through the script. But how would the inputs work? I couldn't type them by hand because it would interrupt the real-time flow of what I was saying. But the obvious thing is just to "autotype" them directly into a notebook.

But how should all this be orchestrated? I start from a script:

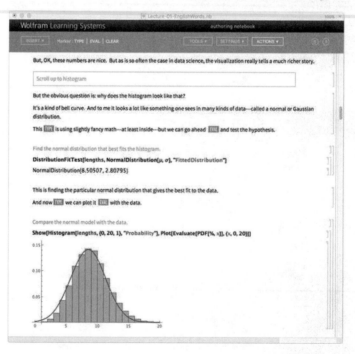

Then I press Generate Recording Configuration. Immediately a title screen comes up in one area of my screen, and I set up my screen-recording system to record from this area. Elsewhere on my screen is the script. But what about the controls? Well, they're just another Wolfram Notebook, that happens to act as a palette containing buttons:

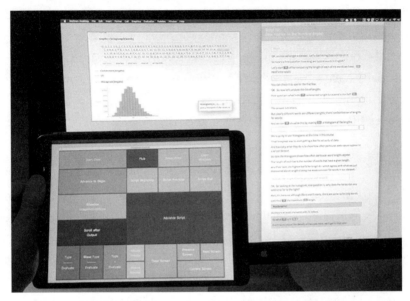

But how can I actually operate this palette? I can't use my mouse, because then I'd take focus away from the notebook that's being screen recorded. So the idea that I had is to put the palette on an extended desktop, that happens to be being displayed on an iPad. So then to "perform" the script, I just press buttons on the palette.

There's a big Advance Script button. And let's say I've read to a point in the script where I need to type something into the notebook. If I want to simulate actual typing I press Slow Type. This will enter the input character-at-a-time into the notebook (yes, we measured the inter-key delay distribution for human typing, and simulate it). After a while it gets annoying to see all that slow typing. So then I just use the Type button, which copies the whole input immediately into the notebook. If I press the button again, it'll perform its second action: Evaluate. And that's the equivalent of pressing Shift+Enter in the notebook (with some optional extra explanatory popups suitable for the video).

I could go on about other tools I've had built using the Wolfram Language, but this gives a flavor. But what do I use that isn't Wolfram Language? Well, I use a web browser, and things that can be reached through it. Still, quite often, I'm just going to the Wolfram Cloud, and for example viewing or using cloud notebooks there.

Sometimes I'll use our public Wolfram Cloud. But more often I'll use a private Wolfram Cloud. The agendas for most of our internal meetings are notebooks that are hosted on our internal Wolfram Cloud. I also personally have a local private Wolfram Cloud running, that I host an increasing number of applications on.

Here's the dock on my computer as of right now:

It's got a filesystem browser; it's got an email client; it's got three web browsers (yes, I like to test our stuff on multiple browsers). Then I've got a calendar client. Next is the client for our VoIP phone system (right now I'm alternating between using this, and using audio along with our screensharing system). Then, yes, at least right now I have a music app. I have to say it's rather rare that my day gives me a chance to listen to music. Probably the main time when I end up doing it is when I'm very behind on email, and need something to cheer me up as I grind through thousands of messages. As soon as I'm actually writing anything nontrivial, though, I have to pause the music, or I can't concentrate. (And I have to find music without vocals—because I've noticed I can't read at full speed if I'm hearing vocals.)

Sometimes I'll end up launching a standard word processor, spreadsheet, etc. app because I'm opening a document associated with one of these apps. But I have to admit that in all these years I've essentially never authored a document from scratch with any of these apps; I end up just using technology of ours instead.

Occasionally I'll open a terminal window, and directly use operating system commands. But this is becoming less and less common—because more and more I'm just using the Wolfram Language as my "super shell." (And, yes, it's incredibly convenient to store and edit commands in a notebook, and to instantly be able to produce graphical and structured output.)

As I write this, I realize a little optimization I haven't yet made. On my personal homepage there are some links that do fairly complex things. One, for example, initiates the process for me doing an unscheduled livestream: it messages our 24/7 system monitoring team so they can take my feed, broadcast it, and monitor responses. But I realize that I still have quite a few custom operating system commands, that do things like update from the source code repository, that I type into a terminal window. I need to set these up in my private cloud, so I can just have links on my personal homepage that run Wolfram Language code for these commands. (To be fair, some of these commands are very old; for example, my fmail command that sends a mail message in the future, was written nearly 30 years ago.)

But, OK, if I look at my dock of apps, there's a definite preponderance of Spikey ones. But why, for example, do I need three identical standard Spikeys? They're all the Wolfram Desktop app. But there are three versions of it. The first one is our latest distributed version. The second one is our latest internal version, normally updated every day. And the third one (which is in white) is our "prototype build," also updated every day, but with lots of "bleeding edge" features that aren't ready to go into serious testing.

It requires surprisingly fancy operating system footwork to get these different versions installed every night, and to correctly register document types with them. But it's very important to my personal workflow. Typically I'll use the latest internal version (and, yes, I have a directory with many previous versions too), but occasionally, say for some particular meeting, I'll try out the prototype build, or I'll revert to the released build, because things are broken. (Dealing with multiple versions is one of those things that's easier in the cloud—and we have a whole array of different configurations running in internal private clouds, with all sorts of combinations of kernel, front end, and other versions.)

When I give talks and so on, I almost always use the latest internal version. I find that livecoding in front of an audience is a great way to find bugs—even if it sometimes makes me have to explain, as I put it, the "disease of the software company CEO": to always want to be running the latest version, even if it hasn't been seriously tested and was built the night before.

Archiving & Searching

A critical part of my personal infrastructure is something that in effect dramatically extends my personal memory: my "metasearcher." At the top of my personal homepage is a search box. Type in something like "rhinoceros elephant" and I'll immediately find every email I've sent or received in the past 30 years in which that's appeared, as well as every file on my machine, and every paper document in my archives:

To me it's extremely convenient to have a count of the messages by year; it often helps me remember the history or story behind whatever I'm asking. (In this case, I can see a peak in 2008, which is when we were getting ready to launch Wolfram|Alpha—and I was working on data about lots of kinds of things, including species.)

Of course, a critical piece of making my metasearcher work is that I've stored so much stuff. For example, I actually have all the 815,000 or so emails that I've written in the past 30 years, and all the 2.3 million (mostly non-spam) ones I've received. And, yes, it helps tremendously that I've had a company with organized IT infrastructure etc. for the past 32 years.

But email, of course, has the nice feature that it's "born digital." What about things that were, for example, originally on paper? Well, I have been something of an "informational packrat" for most of my life. And in fact I've been pretty consistently keeping documents back to when I started elementary school in 1968. They've been re-boxed three times since then, and now the main ones are stored like this:

(I also have file folder storage for documents on people, organizations, events, projects, and topics.) My rate of producing paper documents increased through about 1984, then decayed quite rapidly, as I went more digital. Altogether I have about a quarter million pages of primary non-bulk-printed documents—mostly from the earlier parts of my life.

About 15 years ago I decided I needed to make these searchable, so I initiated the project of scanning all of them. Most of the documents are one or a few pages in length, so they can't be handled by an automatic feeder—and so we set up a rig with a high-resolution camera (and in those days it needed flash). It took several person-years of work, but eventually all the documents were scanned.

We automatically cropped and white-balanced them (using Wolfram Language image processing), then OCR'ed them, and put the OCR'ed text as a transparent layer into the scanned image. If I now search for "rhinoceros" I find eight documents in my archive. Perhaps not surprisingly given that search term, they're a bit random, including for example the issue of my elementary school magazine from Easter 1971.

OCR works on printed text. But what about handwritten text? Correspondence, even if it's handwritten, usually at least comes on printed letterhead. But I have many pages of handwritten notes with basically nothing printed on them. Recognizing handwriting purely from images (without the time series of strokes) is still beyond current technology, but I'm hoping that our neural-net-based machine learning systems will soon be able to tackle it. (Conveniently, I've got quite a few documents where I have both my handwritten draft, and a typed version, so I'm hoping to have a training set for at least my personal handwriting.)

But even though I can't search for handwritten material, I can often find it just by "looking in the right box." My primary scanned documents are organized into 140 or so boxes, each covering a major period or project in my life. And for each box, I can pull up thumbnails of pages, grouped into documents. So, for example, here are school geography notes from when I was 11 years old, together with the text of a speech I gave:

I have to say that pretty much whenever I start looking through my scanned documents from decades ago I end up finding something unexpected and interesting, that very often teaches me something about myself, and about how I ended up developing in some particular direction.

It may be something fairly specific to my life, and the fact that I've worked on building long-term things, as well as that I've kept in touch with a large number of people over a long period of time, but I'm amazed by the amount of even quite ancient personal history that I seem to encounter practically every day. Some person or some organization will contact me, and I'll look back at information about interactions I had with them 35 years ago. Or I'll be thinking about something, and I'll vaguely remember that I worked on something similar 25 years ago, and look back at what I did. I happen to have a pretty good memory, but when I actually look at material from the past I'm always amazed at how many details I've personally forgotten.

I first got my metasearcher set up nearly 30 years ago. The current version is based on Wolfram Language CreateSearchIndex/TextSearch functionality, running on my personal private cloud. It's using UpdateSearchIndex to update every few minutes. The metasearcher also "federates in" results from APIs for searching our corporate websites and databases.

But not everything I want can readily be found by search. And another mechanism I have for finding things is my "personal timeline." I've been meaning for ages to extend this, but right now it basically just contains information on my external events, about 40 of them per year. And the most important part is typically my "personal trip report," which I meticulously write, if at all possible within 24 hours.

Usually the trip report is just text (or at least, text structured in a notebook). But when I go to events like trade shows I typically bring a tiny camera with me, that takes a picture every half-minute. If I'm wearing one of those lanyard name tags I'll typically clip the camera on the top of the name tag, among other things putting it at an ideal height to capture name tags of people I meet. When I write my personal trip report I'll typically review the pictures, and sometimes copy a few into my trip notebook.

But even with all my various current sources of archival material (which now include chat messages, livestreams, etc.), email still remains the most important. Years ago I decided to make it easy for people to find an email address for me. My calculation was that if someone wants to reach me, then in modern times they'll eventually find a way to do it, but if it's easy for them just to send email, that's how they'll contact me. And, yes, having my email address out there means I get lots of email from people I don't know around the world. Some of it is admittedly strange, but a lot is interesting. I try to look at all of it, but it's also sent to a request tracker system, so my staff can make sure important things get handled. (It is sometimes a little odd for people to see request tracker ticket metadata like SWCOR #669140 in email subject lines, but I figure it's a small price to pay for making sure the email is actually responded to.)

I might mention that for decades email has been the primary means of communication inside our (geographically distributed) company. Yes, we have project management, source control, CRM, and other systems, as well as chat. But at least for the parts of the company that I interact with, email is overwhelmingly dominant. Sometimes it's individual emails being sent between people. Sometimes it's email groups.

It's been a running joke for a long time that we have more email groups than employees. But we've been careful to organize the groups, for example identifying different types by prefixes to their names (t- is a mailing list for a project team, d- a mailing list for a department, l- a more open mailing list, r- a mailing list for automated reports, q- a request list, etc.) And for me at least this makes it plausible to remember what the right list is for some mail I want to send out.

Databases of People & Things

I know a lot of people, from many different parts of my life. Back in the 1980s I used to just keep a list of them in a text file (before then it was a handwritten address book). But by the 1990s I decided I needed to have a more systematic database for myself—and created what I started calling pBase. In recent years the original technology of pBase began to seem quite paleolithic, but I now have a modern implementation using the Wolfram Language running in my personal private cloud.

It's all quite nice. I can search for people by name or attributes, or—if I'm for example going to be visiting somewhere—I can just have pBase show me a map of our latest information about who's nearby:

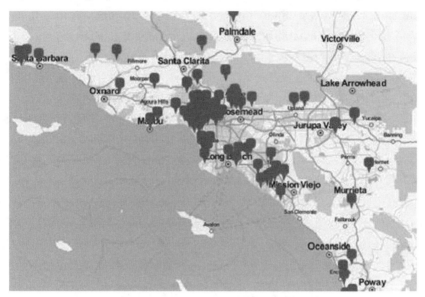

How does pBase relate to social networks? I've had a Facebook account for a long time, but it's poorly curated, and always seems to ride at the maximum number of possible friends. LinkedIn I take much more seriously, and make a point of adding people only if I've actually talked to them (I currently have 3005 connections, so, yes, I've talked to quite a few people).

It's very convenient that every so often I can download data from my LinkedIn account via ServiceExecute to update what's in pBase. But LinkedIn captures only a fraction of people I know. It doesn't include many of my more prominent friends and acquaintances, as well as most academics, many students, etc.

Eventually I'll probably get pBase developed more, and perhaps make the technology generally available. But within our company, there's already a system that illustrates some potential aspirations: our internal company directory—which is running in our internal private cloud, and basically uses Wolfram|Alpha-style natural language understanding to let one ask natural language questions.

I might mention in addition to our company directory, we also maintain another database that I, at least, find very useful, particularly when I'm trying to figure out who might know the answer to some unusual question, or who we might tap for some new project. We call it our Who Knows What database. And for each person it gives a profile of experience and interests. Here's the entry for me (and the source with the question details is online*):

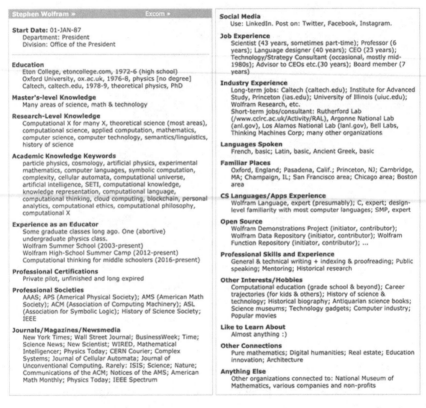

In terms of personal databases, another useful one for me is the database of books I own. I haven't been buying too many books in the past decade or so, but before then I accumulated a library of about 6000 volumes, and it's not uncommon—particularly when I'm doing more historically oriented research—that I'll want to consult quite a few of

---

* www.wolfr.am/WhoKnowsWhatForm

them. But how should they be organized? "Big" classification schemes like Dewey Decimal or Library of Congress are overkill, and don't do a great job of matching my personal "cognitive map" of topics.

Like my filesystem folders, or my physical folders of papers, I've found the best scheme is to put the books into fairly broad categories—small enough in number that I can spatially remember where they are in my library. But how should books be arranged within a category?

Well, here I get to tell a cautionary tale (that my wife regularly uses as an example) of what can go wrong in my kind of approach. Always liking to understand the historical progression of ideas, I thought it would be nice to be able to browse a category of books on a shelf in historical order (say, by first publication date). But this makes it difficult to find a specific book, or, for example, to reshelve it. (It would be easier if books had their publication dates printed on their spines. But they don't.)

About 20 years ago I was preparing to move all my books to a new location, with different lengths of shelves. And I had the issue of trying to map out how to arrange book categories on the new shelves ("how many linear feet is quantum field theory and where can it fit in?") So I thought: "Why not just measure the width of each book, and while I'm at it also measure its height and its color?" Because my idea was that then I could make a graphic of each shelf, with books shown with realistic widths and colors, then put an arrow in the graphic to indicate the location (easily identified visually from "landmarks" of other books) of a particular book.

I got a colorimeter (it was before ubiquitous digital cameras) and started having the measurements made. But it turned out to be vastly more labor-intensive than expected, and, needless to say, didn't get finished before the books had to be moved. Meanwhile, the day the books were moved, it was noticed that the packing boxes fit more books if one didn't just take a single slab of books off a shelf, but instead put other books around the edges.

The result was that 5100 books arrived, basically scrambled into random order. It took three days to sort them. And at this point, I decided just to keep things simpler, and alphabetize by author in each category. And this certainly works fine in finding books. But one result of my big book inventory project is that I do now have a nice, computable version of at least all the books connected to writing *A New Kind of Science*, and it's actually in the Wolfram Data Repository:

In[ ]:= `ResourceData["Books in Stephen Wolfram's Library"]`

Authors	Title
Aarts, Emile & Jan Korst	Simulated Annealing and Bo
Abelson, Harold & Andrea A. diSessa	Turtle Geometry: The Comp
Abelson, Harold, Gerald Jay Sussman & Julie Sussman	Structure and Interpretation
Abraham, Ralph & Christopher D. Shaw	Dynamics – The Geometry o
Abraham, Ralph & Jerrold E. Marsden	Foundations of Mechanics. S
Abraham, Ralph H. & Christopher D. Shaw	Dynamics – The Geometry o
Abraham, Ralph H. & Christopher D. Shaw	Dynamics – The Geometry o
Abramowitz, Milton & Irene Stegun	Handbook of Mathematical F
Abrikosov, A.A., L.P. Gorkov & I.E. Dzyaloshinski	Methods of Quantum Field T
Achacoso, Theodore B. & William S. Yamamoto. Editors	AY's Neuroanatomy of C. ele

Personal Analytics

In 2012 [last chapter] I wrote a piece about personal analytics and the data I've collected on myself. Back then I had about a third of a million emails in my archive; now it's half a million more.

I have systems that keep all sorts of data, including every keystroke I type, every step I take, and what my computer screen looks like every minute (sadly, the movie of this is very dull). I also have a whole variety of medical and environmental sensors, as well as data from devices and systems that I interact with.

It's interesting every so often to pick up those Wolfram Data Drop databins and use them to do some data science on my life. And, yes, in broad terms I find that I am extremely consistent and habitual— yet every day there are different things that happen, that make my "productivity" (as measured in a variety of ways) bounce around, often seemingly randomly.

But one thing about collecting all this data is that I can use it to create dashboards, and these I find useful every single day. For example, running in my private cloud is a monitoring system for my email:

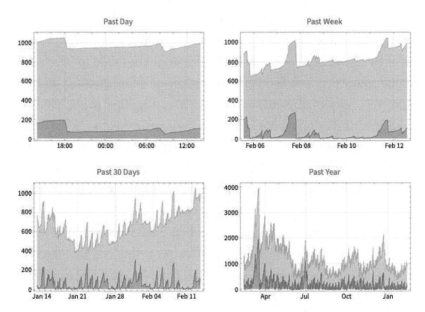

The top curve is my total number of pending email messages; the bottom is the number I haven't even opened yet. These curves are pretty sensitive to all kinds of features of my life, and for example when I'm intensely working on some project, I'll often see my email "go to seed" for a little while. But somehow in trying to pace myself and decide when I can do what, I find this email dashboard very helpful.

It's also helpful that every day I get emails reporting on the previous day. How many keystrokes did I type, and in what applications? What files did I create? How many steps did I take? And so on.

I keep all kinds of health and medical data on myself too, and have done so for a long time. It's always great to have started measuring something a long time ago, so one can plot a several-decade time series and see if anything's changed. And, actually, the thing I've noticed is that often my value (say blood level) for something has remained numerically essentially the same for years—but many of the "normal ranges" quoted by labs have bounced all over the place. (Realistically this isn't helped by labs inferring normal ranges from their particular observed populations, etc.)

I got my whole genome sequenced in 2010. And although I haven't learned anything dramatic from it, it certainly helps me feel connected to genomic research when I can see some SNP variant mentioned in a paper, and I can immediately go look to see if I have it. (With all the various vicissitudes of strands, orientations and build numbers, I tend to stick to first principles, and just look for flanking sequences with StringPosition.)

Like so many of the things I've described in this piece, what has worked for me in doing personal analytics is to do what's easy to do. I've never yet quite solved the problem, for example, of recording what I eat (our image identification isn't yet quite good enough, and even made-just-for-me apps to enter food have always seemed a bit too onerous). But whenever I have a system that just operates automatically, that's when I successfully collect good personal analytics data. And having dashboards and daily emails helps both in providing ongoing feedback, and in being able to check if something's gone wrong with the system.

The Path Ahead

I've described—in arguably quite nerdy detail—how some of my personal technology infrastructure is set up. It's always changing, and I'm always trying to update it—and for example I seem to end up with lots of bins of things I'm not using anymore (yes, I get almost every "interesting" new device or gadget that I find out about):

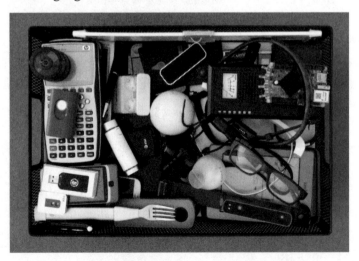

But although things like devices change, I've found that the organizational principles for my infrastructure have remained surprisingly constant, just gradually getting more and more polished. And—at least when they're based on our very stable Wolfram Language system—I've found that the same is true for the software systems I've had built to implement them.

What of the future? Some things will certainly get upticked. I realized while writing this chapter that I can now upgrade to 4k monitors (or higher) without affecting screensharing (the feed is automatically downsampled). Before too long maybe I'll be using AR to annotate my environment in real time. Maybe eventually I'll have some way to do XR-based as-if-in-person videoconferencing. Maybe—as I've been assuming will be possible for 40+ years—I'll finally be able to type faster using something like EEG. And so on.

But the more important changes will be in having better-developed, and more automated, workflows. In time I expect it'll be possible to use our machine learning tools to do automatic "computational history," for example assembling a useful and appropriately clustered timeline of things I've done, say in a particular area.

In my efforts at historical research, I've had occasion to use lots of archives of people and organizations. There's usually a certain amount of indexing and tagging that's been done. (Who is that letter to and from? When was it written? What are its keywords? Where was it filed? And so on.) But things tend to be very granular, and it's usually hard work to determine the overall arc of what happened.

My first goal is to make all the material I personally have useful for myself. But I'm thinking of soon starting to open up some of the older material for other people to see. And I'm studying how—in modern times, with all the cloud infrastructure, machine learning, visualization, computational documents, etc. that we have—I can build the best possible system for presenting and exploring archives.

As I think about my day, I ask myself what aspects of it aren't well optimized. A lot of it actually comes down to things like email processing, and time spent for example actually responding to questions. Now, of course, I've spent lots of effort to try to structure things so as many questions as possible become self-answering, or can be addressed with technology and automation that we've built. And, in my role as CEO, I also try hard to delegate to other people whenever I can.

But there's still plenty left. And I certainly wonder whether with all the technology we now have, more could be automated, or delegated to machines. Perhaps all that data I've collected on myself will one day let one basically just built a "bot of me". Having seen so many of my emails—and being able to look at all my files and personal analytics—maybe it's actually possible to predict how I'd respond to any particular question.

We're not there yet. But it will be an interesting moment when a machine can, for example, have three ideas about how to respond to something, and then show me drafts that I can just pick from and approve. The overall question of what direction I want to go in will almost by definition have to stay with me, but the details of how to get there I'm hoping can increasingly be automated.

A Precociousness Record (Almost) Broken

June 1, 2011

I got started with science quite early in my life... with the result that I got my PhD (at Caltech, in physics) when I was 20 years old. Last weekend a young woman named Catherine Beni (whom I had met quite a few years ago) sent me mail saying she had just received her PhD from Caltech (in applied math)—also at the age of 20.

Needless to say, we were both curious who had the record for youngest Caltech PhD. Catherine said she was 20 years, 2 months, and 12 days old when she did her PhD defense. Well, I knew I'd finished my PhD in November 1979—and I was born August 29, 1959. So that would also have made me around 20 years and 2 months old.

I quickly searched the OCR'ed archive that I have of my paper documents, and found this:

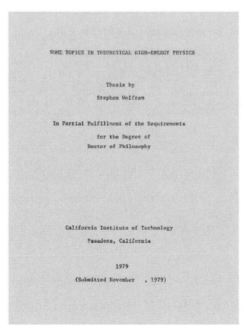

The month was confirmed, but frustratingly, no day was filled in. But then I remembered something about my PhD defense (the little talk that people give to officially get their theses signed off). In the middle of it, I was having a rather spirited discussion (about the second law of thermodynamics) with Richard Feynman, and suddenly the room started shaking—there was a minor earthquake.

Well, now we have Wolfram|Alpha. So I type in "earthquakes at Caltech in Nov. 1979," and out comes:

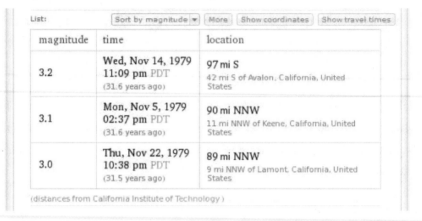

There it is! The only possible date for an afternoon PhD defense is November 5, 1979. So now I can compute my age:

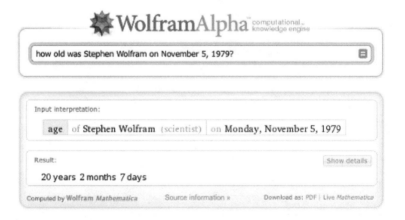

"20 years 2 months 7 days." My 30-year record for youngest Caltech PhD is preserved, by less than a week.

Well, at least sort of. Perhaps the correct official date for "getting a PhD" is the graduation ceremony. And by that measure, Catherine Beni is the winner by almost six months! (Perhaps more: I never actually went to my graduation ceremony, though the certificate did eventually arrive in the mail.)

(I also wondered about the precise time span in days: for me from birth to PhD defense it was 7373 days—and for Catherine the months line up so that it was exactly five days more: 7378 days.)

Whatever the details, Catherine and I are now thinking of starting a curious little club for "low-age PhDs." If the cutoff age is 21, I think there must be a decent number of potential members. Certainly there's Ruth Lawrence, who got her PhD in math at Oxford in 1989 at the age of 17. And then there's Harvey Friedman, who got his PhD in math at MIT in 1967 at 18. And Norbert Wiener, who got his PhD in math at Harvard in 1912 at 18. I'm guessing there are perhaps a dozen other legitimate examples, mostly in math and closely related areas. (So far, I'm the only non-math example I know; quite likely I even have the global record for youngest physics PhD.)

What do I think about precociousness? Over the years, a lot of people have come to me with it, so I've seen quite a bit of it. And if one ignores "precociousness for its own sake" (of which there's an increasing amount these days), what's left is a pretty interesting collection of stories. I would say that perhaps half of them have impressive or at least happy outcomes; the other half do not. My guess is that many of the better outcomes are associated with people who have good early judgment, as well as skill.

For me, at least, precociousness was a huge win. Because it allowed me to launch into adult life early—before whatever enthusiasm and originality I had was ground down by years of structured education.

As it happens, I didn't start off thinking of myself as precocious. I was a top student at top schools in England, but I didn't pay any attention to that. I got interested in physics when I was about 10, and just read more and more about it, and then started doing research about it, writing about it, and eventually publishing papers about it. I pretty much didn't talk about my physics research with anyone. But because I'd taught myself all sorts of fancy techniques and so on, I started being able to do school-level physics very well.

So when I was 16 I left high school (Eton) and went to college (Oxford). (In between I had a job doing theoretical physics at a British government lab.) I didn't last long in college, and never got a degree. But by the time I left I'd published quite a few physics papers (even including some that I still consider quite good)—and I went straight to graduate school at Caltech.

It would have been easy for me to "get a physics PhD while I was still a teenager," but at the time I wasn't thinking of things like that. And so I ended up getting my PhD, as I now know, when I was 20 years, 2 months, and 7 days old.

When I was in high school, people kept on telling me that if I accelerated things as I ended up doing, I would somehow have terrible trouble. "Social difficulties" or something, they said. Well, I'm happy to say that none of that terrible trouble ever materialized.

To be fair, a certain amount of toughness was required on my part at various stages. But the main effect of the prediction of trouble was that it caused me to take longer than it should have to realize that precociousness in science is not incompatible with more obviously worldly things, like running companies.

When I was young, it was fun being precocious, and I think I was lucky that I taught myself so much in my early years. Because somehow it gave me the confidence to believe that I could teach myself almost anything. And in the 31 years since I got my PhD, I've been learning subject after subject, in a sense always simulating my youthful precociousness attitude: "just because other people seem to think this is hard doesn't mean I can't figure it out."

Just in terms of the raw passage of years, being precocious has let me get more done: I've been able to spend more time learning things, and more time doing projects. It's a little weird these days, seeing so many of my "contemporaries" from my early days in physics get to the end of their careers. Because I still feel like I did back when I was being a precocious physics kid: there are so many wonderful things to do, and I'm just getting started...

A Speech for (High School) Graduates

June 9, 2014

Given at the 2014 graduation event for Stanford Online High School:

You know, as it happens, I myself never officially graduated from high school, and this is actually the first high school graduation I've ever been to.

It's been fun over the past three years—from a suitable parental distance of course—to see my daughter's experiences at OHS. One day I'm sure everyone will know about online high schools—but you'll be able to say, "Yes, I was there when that way of doing such-and-such a thing was first invented—at OHS."

It's great to see the OHS community—and to see so many long-term connections being formed independent of geography. And it's also wonderful to see students with such a remarkable diversity of unique stories.

Of course, for the graduates here today, this is the beginning of a new chapter in their stories.

I suspect some of you already have very definite life plans. Many are still exploring. It's worth remembering that there's no "one right answer" to life. Different people are amazingly different in what they'll consider an "'A' in life." I think the first challenge is always to understand what you really like. Then you've got to know what's out there to do in the world. And then you've got to solve the puzzle of fitting the two together.

Maybe you'll discover there's a niche that already exists; maybe you'll have to create one.

I've always been interested in trajectories of people's lives, and one thing I've noticed is that after some great direction has emerged in someone's life, one can almost always look back and see the seeds of it very early.

Like I was recently a bit shocked actually to find some things I did when I was 12 years old—about systematizing knowledge and data—and to realize that what I was trying to do was incredibly similar to Wolfram|Alpha. And then to realize that my tendency to invent projects and organize other kids to help do them was awfully like leading an entrepreneurial company.

You know, it's funny how things can play out. Back when I was a kid I was really interested in physics. And to do physics you have to do a lot of math calculations. Which I found really boring, and wasn't very good at.

So what did I do? Well, I figured out that even though I might not be good at these calculations, I could make a computer be good at them. And needless to say, that's what I did—and through a pretty straight path, that's what brought the world Mathematica and Wolfram|Alpha.

You know, another thing was that when I was a kid I always had a hard time getting myself to do exercises from textbooks. I kept on thinking to myself, "Why am I doing this exercise when zillions of other people have already done it? Why don't I do something different, that's new, and mine?"

People might think: that must be really hard. But it's not. It's just that you have to learn not just about how to do stuff, but also about how to figure out what stuff to do. And actually one thing I've noticed is that in almost every area, the people who go furthest are not the ones with the best technical skills, but the ones who have the best strategy for figuring out what to do.

But I have to say that for me it's just incredibly fun inventing new stuff—and that's pretty much what I've spent my life doing.

I think most people don't really internalize enough how stuff in our world gets made. I mean, everything we have in our civilization—our technology, our ways of doing things, whatever—had to be invented. It had to start with some person somewhere—maybe like you—having an idea. And then that idea got turned into reality.

It's a wonderful thing going from nothing but an idea, to something real in the world. For me, that's my favorite thing to do. And I've been fortunate enough to do that with a number of big projects, alternating between science, technology, and business. At some level, my projects might look very different: building a new kind of science, creating a computer language, encoding the world's knowledge in computational form.

But it turns out that at some level they're really all the same. They're all about taking some complicated area, drilling down to the essence of it, then doing a big project to build up to something that's useful in the world.

And when you think about what it is you really like, and what you're really good at, it's important to be thematic. Maybe you like math. But why? Is it the definiteness? Problem solving? Elegance? Even at OHS you only get to learn about certain specific subjects. So to understand yourself, you have to take your reactions to them, and generalize—figure out the overall theme.

You know, something I've learned is that the more different areas I know about, the better. When I was a kid I learned Latin and Greek—and I was always complaining that they'd never be useful. But then I grew up—and had to make up names for products and things. And actually for years a big part of what I've done every day is to take ideas from very different areas that I've learned about—and bring them together to make new ideas.

One thing, if you want to do this, is that you really have to keep all those things you've learned at your fingertips. History of science can't stay in a history of science class. It has to inform that clever social media idea you have, or that great new policy direction you come up with, or that artistic creation you're making, or whatever. The real payoff comes not from doing well in the class, but from internalizing that way of thinking or that knowledge so it becomes part of you.

You know, as you think about what to do in the world, it's worth remembering that some of the very best areas are ones that almost nobody's heard about yet—and there certainly aren't classes about. But if you get into one of those new areas, it's great—because there's still all this basic ground-floor stuff to do there, and as the area grows, you get propelled by that.

I've been pretty lucky in that regard. Because early in life I got really interested in computation, and in the computational way of thinking about things. And I think it's becoming clear that computation is really the single most important idea that's emerged in the past century. And that even after all the technology that's been built with it, we're only just beginning to see its true significance.

And today, you just have to prepend the word "computational" to almost any existing field to get something that's an exciting growth direction: computational law, computational medicine, computational archaeology, computational philosophy, computational photography, whatever.

And yes, to be able to do all this stuff, you have to get familiar with the computational way of thinking, and with things like programming. That's going to be an increasingly important literacy skill. And I have to say that in general, even more valuable than learning the content of specific fields is to learn general approaches and tools—and keep up to date with them.

It's not for everybody, but I myself happen to have spent a lot of time actually building tools. And for me the most powerful thing has been being able to build a tower of tools, then to use them to figure things out, then to use those things to go on and build more tools. And I've been fortunate enough to be able to go on doing that for more than 30 years now.

You know, it's always an interesting judgment call when to go on in a life direction you're already going, and when to branch into something new, or to chase some new opportunity. For myself, I try to maintain a portfolio—continuing to build on what I've done, but also always making sure to add new things.

One of the consequences of that is that at any given time, there's always an area where I'm basically a beginner—and just learning. Right now, for example, that happens to be programming education. We've managed to automate a lot of programming—which I think is going to be a pretty big deal in general—but for education it means there's a much broader range of people and places where programming can be taught. But how should it be done? Math and language and areas like that have centuries of education experience to draw on. But with what's now possible with programming education, we've got a completely new situation, that kind of has to be figured out from the ground up. It's always a little scary doing something like that, and I always think, "Maybe this is finally an area I'll never figure out." But somehow if one has the confidence to keep going, it always seems to come together— and it's really satisfying.

You know, when I was a kid I learned some things in school and some things on my own. I was always doing projects about this or that. And somehow I've just kept on doing projects and learning more and more things. You've been exposed to lots of interesting things at OHS. Make sure you expose yourselves to lots more things in college or wherever you're going next. And don't forget to do projects—to do things that are really yours, and that people can look at and really get a sense of you from.

And don't just learn stuff. Keep thinking about strategy too. Keep trying to solve the puzzle of what your best niche is. You might find it or you might have to create it. But there will be something great out there for you. And never assume that the world won't let you get to it. It's all part of the puzzle to solve. And the seeds are already there in who you are; you just have to find them, nurture them, and keep pushing to let them grow as each chapter of your story unfolds...